T0313343

Reducing Inequalities Towards Sustainable Development Goals: Multilevel Approach

RIVER PUBLISHERS SERIES IN CHEMICAL, ENVIRONMENTAL, AND ENERGY ENGINEERING

Series Editors

ALIREZA BAZARGAN
NVCo and University of Tehran
Iran

MEDANI P. BHANDARI
Akamai University, USA
Sumy State University, Ukraine
Atlantic State Legal Foundation, NY, USA

HANNA SHVINDINA
Sumy State University, Ukraine

Indexing: All books published in this series are submitted to the Web of Science Book Citation Index (BkCI), to SCOPUS, to CrossRef and to Google Scholar for evaluation and indexing.

The "River Publishers Series in Chemical, Environmental, and Energy Engineering" is a series of comprehensive academic and professional books which focus on Environmental and Energy Engineering subjects. The series focuses on topics ranging from theory to policy and technology to applications.

Books published in the series include research monographs, edited volumes, handbooks and textbooks. The books provide professionals, researchers, educators, and advanced students in the field with an invaluable insight into the latest research and developments.

Topics covered in the series include, but are by no means restricted to the following:

- Energy and Energy Policy
- Chemical Engineering
- Water Management
- Sustainable Development
- Climate Change Mitigation
- Environmental Engineering
- Environmental System Monitoring and Analysis
- Sustainability: Greening the World Economy

For a list of other books in this series, visit www.riverpublishers.com

Reducing Inequalities Towards Sustainable Development Goals: Multilevel Approach

Editors

Medani P. Bhandari

Akamai University, Hawaii, USA
Sumy State University, Ukraine

Shvindina Hanna

Sumy State University, Ukraine

LONDON AND NEW YORK

Published 2019 by River Publishers
River Publishers
Alsbjergvej 10, 9260 Gistrup, Denmark
www.riverpublishers.com

Distributed exclusively by Routledge
4 Park Square, Milton Park, Abingdon, Oxon OX14 4RN
605 Third Avenue, New York, NY 10017, USA

Reducing Inequalities Towards Sustainable Development Goals: Multilevel Approach / by Medani P. Bhandari, Shvindina Hanna.

Routledge is an imprint of the Taylor & Francis Group, an informa business

ISBN 978-87-7022-126-9 (print)

While every effort is made to provide dependable information, the publisher, authors, and editors cannot be held responsible for any errors or omissions.

"Dasha ... She is associated with one word – WISDOM. Wisdom that includes critical thinking, deep knowledge, and most important, deep feelings and emotions. This cocktail, a complicated combination, made her real. She was real in love, in friendship, in her attitude to work, people, and projects she was involved into. She got unshakable wish to help people to see new opportunities and get them. She was generous in sharing these opportunities. She helped a lot of people to get scholarships or projects abroad that had changed lives. Because of Dasha I've started the path to myself. I learned how to slow down and be real, not "liked". I learned to appreciate life for real and accept the challenges as the reserve for growth and strength. Today I know how to transform the pain into gratitude and love. And the wisdom of that, at first sight fragile, girl will live inside of me ... Her life force was enough to change the worlds of many of us ... "

Alina Poznanska, Sumy State University Alumni, Head of NGO "Active Citizen City" and Co-founder of "POZNANSKIteam"

"Dasha was one of those people who would light up a room with her smile. She was that rare package, a warm, smart, caring a funny person. I was privileged to have her as a friend and proud to have her as a co-creator of the Active Citizens environment in Ukraine, since it's earliest days. She is missed"

Mike Waldron, Chief Executive and Founder of Inter-Pares Associates Ltd., Chairman of the Board of NGO "Global Tolerance", Trainer of Active Citizens Programme

"I always admired Daryna secretly. She was so light, so scintillating, such a star. She managed to bring positive vibes into everything, whatever she was doing. She lighted people up, she lighted me up too. Only she could grab your hand and told you about a new life opportunity for YOU PERSONALLY and change your life for good. Without her I would not have my Erasmus Scholarship, and who knows, other projects and victories. But definitely, without her, this book would not be possible. And when I am looking back, I understand that we were just lucky to have such a sunny shining person in our lives"

Shvindina Hanna, Educator, Sumy State University, Associate Professor and CEO at NGO "Lifelong Learning Centre"

We all authors and editors, dedicate this volume to Daryna Leus (Boronos), (our young dearest colleague, daughter, friends of everyone, mother, an excellent housewife, lecturer – whose moto of life was to contribute to the society in fullest with all possible ways. Her departure to eternity in such young age has been very painful to all of us. We have been noticing hollowness to all of us- especially in the eyes of her father Prof. Volodymyr Boronos, (no one can understand that pain), her child, her students and we all listed in the acknowledgements). We never forget you Daryna, we dedicate this book to you, and pray for your peace of soul.

On behave of all authors and editorial team – Medani P. Bhandari, PhD Professor of Sustainability, Enterprenurship and Innovation

Contents

Foreword

Based in this text, the goal of significant reduction of economic and social inequalities in developing countries, remains unachieved. Amelioration of human suffering in the third world continues as an important goal for the United Nations. The authors of this important book address these issues and lay out the concerns in an effective manner, with an important focus upon equivalency of the global moral standard. The title of the book, "Reducing Inequalities Towards Sustainable Development Goals: Multilevel Approach" must be viewed among the premier missions of the global community. Prof. Dr. Medani P. Bhandari and his colleague, Associate Prof. Shvindina Hanna, in their editing, have addressed a path forward toward the final cure for human inequality.

The hope of the book parallels the philosophies of many of the great leaders of the international humanitarian movement. For instance, the American human rights hero, Dr. Martin Luther King, Jr. once stated: *Everybody can be great ... because everybody can serve ... You only need a heart full of grace and a soul generated by love.* He said: *The time is always right ... to do what is right.* The great Filipino Benigno Ninoy Aquino once said: *We should not depend on one man, we should depend on all of us. All of us in the cause for freedom must stand up now and be a leader, and when all of us are leaders, we will expedite the cause of freedom. Overcoming major human challenges is the reality of your world, as remembered in the words of the great Vietnam leader, Ho Chi Minh: "Remember that the storm is a good opportunity for the pine and the cypress to show their strength and their stability. After the rain ... good weather In the wink of an eye, the universe throws off its muddy clothes. Further, Mahatma Gandhi stated: As human beings, our greatness lies not so much in being able to remake the world, rather our greatness rests – in being able to remake ourselves. A man is but the product of his thoughts ... what he thinks he becomes.* Further, he said: *My life is my message ... You must be the change you wish to see in the world.*

These men were great humanitarians because they had a true mission, and they pursued it with every ounce of humanity. As is a foundation of this book, the global community must establish human equality and betterment of the human condition as primary goals for all humanity. World leaders then, must achieve this goal with unity.

Another basis for betterment of the human condition is founded in a premise of the major world religions. The holy teachings of Jesus of Nazareth states: *whatever you want others to do to you, do also to them....* This is widely referred to as the Golden Rule, and internationally, across the major religions, this teaching underscores law and community across much of the human race. For Buddhism, the same teaching states: *Hurt not others in ways that you yourself would find hurtful.* For Hinduism, we find written: *This is the sum of duty; do nothing unto others ... what you would not have them do unto you.* In Islam, it is believed that: *No one of you is a believer until he desires for his brother that which* he desires for himself.

These foundations then support the premise of this book, that the nations united in policy and effort must move toward this goal of global human equality. I thank the authors for pursuing such a grand undertaking

Dr. Douglass Capogrossi,
President, Akamai University, Hilo, Hawaii, USA

Міністерство освіти і науки України   Ministry of Education and Science of Ukraine

СУМСЬКИЙ ДЕРЖАВНИЙ
УНІВЕРСИТЕТ

вул. Римського-Корсакова, 2, м.Суми, 40007
тел.(0542) 64-04-99, тел/факс (0542) 33-40-58
e-mail: kanc@sumdu.edu.ua
sumdu.edu.ua
код ЄДРПОУ 05408289

SUMY STATE
UNIVERSITY

2, Rymsky-Korsakov St. Sumy, Ukraine, 40007
tel.+38 (0542) 64-04-99, tel/fax +38 (0542) 33-40-58
e-mail: kanc@sumdu.edu.ua
sumdu.edu.ua
USREOU code 05408289

№ 53.00/01-02/3032
від 0 3 ЛИП 2019 _____ 20__ р.
на № _____
від _____ 20__ р.

Forward word for Monograph

This edited book - is a combined effort of several faculties of Sumy State University and others mainly based in Ukraine. In 2017, Sumy State University invited Prof. Medani P. Bhandari for a week-long educational program developed by University faculties coordinated by Associate Professor Shvindina Hanna and Daryna Leus (who is unfortunately no more with us). Prof. Bhandari's main discussion and lectures for the faculties and students were how to encourage and engage faculties and doctoral students on research and knowledge production, in a productive way. During one of the meetings with faculties, Daryna initiated the idea to create the edited volumes as a practical example. The project was led by Professor Bhandari who has expertise in the spheres of Social justice and Environmental Protection, Social and Environmental policies, Climate Change and Sustainable development and Associate Professor Hanna Shvindina who enthusiastically contributes into Sustainable Development domain and is considered as one of promising project managers who directs her projects towards a sustainable and better future. The authors of this volume are highly qualified in their respected fields. Each chapter is theory grounded and reveals data-driven results. The book is an example of the wealth of knowledge and fills the knowledge gap in the field of development economics, inequality (gender, states, economic, geographical and social) and the interrelationship with the sustainability agendas. This monograph will be helpful for researchers, students and graduates, practitioners, strategists, public officers and policymakers, agencies and governments. The outcomes of the research may be provided as a part of the courses in a sphere of civic education, training on management, environmental economics and other branches of social science. I would like to express my appreciation to Professor Medani Bhandari and Associate Professor Hanna Shvindina for their leadership in this project. And to all authors for their contribution to the volume.

Rector   Anatoliy Vasylyev

Preface

This book provides factual evidence of increased inequality thorough country specific case studies and comparative studies. Transnational. In general notion, inequality is seen every sphere of human history, socially, politically, economically and environmentally. Inequality of any nature whether it is through social, economic, cultural, political or environmental strata, creates the division on humanity and has been one of the major causes of conflict throughout human civilization.

The major goal of any society is sustained harmony in the society, however, it rarely exists in human sociopolitical history. The divisive element of inequality – is oppression to weak according to strata which can be economic, social, political, cultural or others forms-grounded and influenced by societal circumstances. Theoretically, it is hard to fully pinpoint the root cause of inequality- why oppressed are being oppressed and oppressors have been maintaining the domination. However, social justice theory, social dominance theory and development economics theory provides some basis of understand the nature of inequality. "The goal of social justice is full and equal participation of all groups in a society that is mutually shaped to meet their needs" – However, mostly societies are maintaining their structure based on dominance, therefore, inequality fostering in the society. Development economics is grounded on the basic notion of how poorest of the poor country's economic condition can be improved in holistic way covering market condition, national and international economic policies, its population and labor condition, health status, educational structure and output. However, there are hurdle in implementation of development principles to address the inequality features because they are complicated economically, politically, socially, culturally and uncontrolled/unseen/invisible walls within the societal strata local to international. Despite of many efforts of governments, civil society, media and advocacy, leadership, scholars and all who believe inequality is factor of social division; inequality is increasing locally to internationally. Internationally, United Nation has been pioneering to find the root cause of inequality and find the way to minimize it throughout its history.

United Nation declared four decades (1960–1990) as development decade with the objective of total development primarily in the developing world. In 1990, UN presented Human Development Report 1990, and in 2000 UN declared millennium development goals (2000–2015). However, Goals were only partially achieved. With this experience, UN declared "Transforming our world: the 2030 Agenda for Sustainable Development, which declared 17 Sustainable Development Goals and 169 targets. Goals 10 solely stands for reducing inequality, however, Goals 1, 3, 4, 5, 6, 7, 8, 9, 10, 11, 16 and 17 also linked with inequality issues. In this edited book, authors have tried to find out the current state inequality, through economic point of view (mostly development economics). The results of this books chapter show, despite of so much efforts from all concerned stakeholders, inequality has not been reduced. In fact, mostly since 1980, inequality has been rapid growth in the world (both developing and developing world).

Endorsements

Reducing Inequalities Towards Sustainable Development Goals: Multilevel Approach – is the results of the collaboration of scientists and researchers from USA and Ukraine in their attempt to re-think the current situation and find out solutions for well-known but still unsolved problems. Sustainable development is the contemporary survival ideology and one of the greatest challenges of the modern world and we need to continue moving towards it and uniting our efforts. All authors have tried to find the way of overcoming the economic, environmental and social inequalities that shaped by 17 goals formulated by UN in the 2030 Agenda for Sustainable Development and specified for every country.

Every research touches some specific issue, nevertheless what really united all participants of the book – their aspiration to find out the solutions inter alia institutional, regulation, fiscal, and political. The results of the research presented in the book by different authors and the book in general, edited by Dr. Medani P. Bhandari and Dr. Shvindina Hanna, will be helpful for experts, policymakers, NGO activists and agencies as a basis for understanding the most crucial problems that should be in a focus of our common attention.

Oleksander Romanovskiy
Dr. of Sci., Academician of the Academy of Sciences of the Higher Education System Rector of Ukrainian-American Concordia University, Kyiv, 01601, Ukraine

This book "Reducing Inequalities Towards Sustainable Development Goals: Multilevel Approach" edited by Profs. Medani P. Bhandari and Shivandina, is a product of a scenario which tries to reinterpret the concept of sustainable development goals, conceived decades ago, and relevancy in relation to inequality. Inequality has existed from time immemorial in history and continues reverberating across all continents with intensification of economic liberalization. Consequences of this is innovations and technological advancement which are marshalling swift and subtle changes touching human life and eco-system. Scholars, scientists, economists, philosophers, policy makers and leadership are not sure of ways of managing 'change' at par with the speed of changes occurring. Rather challenging environment. Semblance of uncertainty and concomitant fear is inherent during '*change*' process and this book is one of a kind in this journey. In the immediate past decades, platform economy created unprecedented level of new wealth, unthinkable to traditionalists. This phenomenon is widening income inequality gaps in one side while burgeoning wealth concentration continues to amass with few on the other end. This little understood challenge is unlikely to stop anytime soon for innovations are taking place all over the globe, unlike in the past. This trend is surely going to impact on skills profile, education systems, human habits, including in socio-political and economic behaviors of the future. In this, growing application of artificial intelligence will heavily impact on industrial and manufacturing processes and services sector resulting manifold disruptions all around including in labor force, work, market and income disparity. Perhaps, it is an avoidable reality of the future. No one appears prepared for this '*change*' because there is no known proven clear path or a model that one could emulate in from the past.

This book provides a unique analysis on growing inequalities. To this end, United Nations global initiatives through three development decades (1960–90), Agenda 21 (1991–2000), millennium development goals (2000–15) and Sustainable Development Goals (2015–2030) are noteworthy. Authors have concluded that United Nations efforts have not fully met the goal of significant reduction of economic and social inequalities in developing countries, East European countries and North America. This book depicts a vivid picture of economic inequality. This is a useful thesis for all interested stakeholders including economists, scholars, students, educationists, civil societies, governments, development agencies and United Nations at large. I could not disagree with Prof. Bhandari's conclusion that "the invisible walls are everywhere and "profit first" is the dominant mainstay of the modern growth mantra of development paradigms; an inescapable realism of unfettered wealth creation theme of the era. And his "Bashudaiva Kutumbakkam" is a noble philosophical concept that could be aspiring for global harmony among nation states and communities, and a guide to policy makers for new ways of looking into socio-economic disparities; certainly, a tall order in an environment of economic protectionism and political populism, currently in vogue, and ensuing challenges to foster "live well and let live" as one global fellow community.

Kedar Neupane
Former staff of United Nations High Commissioner for Refugees (UNHCR)
Geneva, Switzerland
14 July 2019

It is my pleasure to write a few words about Dr. Medini P. Bhandari and Shvindina Hanna's work. I have known Dr. Bhandari since 2017. His scholarship, research and in-depth analysis of subjects he handles are remarkably profound and timely for major shifts in policy decisions by all the concerned societies and countries equally. I have read his previous work "Green Web-II: Standards and Perspectives from the IUCN. It is a very profound analysis of justice, equity and sustainable development. In this edited work "Reducing Inequalities Towards Sustainable Development Goals: Multilevel Approach" his analysis is very crucial to fathom the depth of globe marching towards its own destruction. All scholar's contribution in the volumes are His write-up in "The Problems and remarkably maintain very high standards, which adds new knowledge in inequality and Sustainable Development". Every scholar succinctly lays out the fundamental issues of the world today. They are all divides, viz economic, income, race, religion and, human relations as well as geopolitical intolerance. This has given the rise to further dictatorial attitudes to haves in all lever of social strata including governance. Book outlines statistical analysis of the issues in the world today. I admire Dr. Bhandari, Hanna's and all authors for frank and timely pointers that humanity needs to wake up. Otherwise we may be doomed before we see the twenty second century. I congratulate Dr. Medini P. Bhandari, Hanna and their team for their attempt to help mankind rise to the occasion to bring in sanity to its own existence.

Thank you

Bishnu P. Poudel, Ph.D
Professor of International Diplomacy and Relations

It is my great pleasure to write a few words on Reducing Inequalities Towards Sustainable Development Goals: Multilevel Approach, edited by Profs. Medini P. Bhandari and Shvindina Hanna. This literature review-based work on inequality is very informative and highly useful. It addresses the major challenges the human society has faced due to income inequality and disparity. In various chapters, authors have analyzed secondary data on income inequality and disparity and has explained the efforts made by the United Nations to narrow the problems of inequality and disparity to achieve the sustainable development goals and targets. The authors have highlighted how income inequality and disparity create several environmental challenges and these environmental challenges exasperate income inequality and disparity. Authors also used examples to highlight how these income inequality and disparity have created several environmental problems in many developing countries including in the Post-Soviet Ukraine, and these environmental problems in turn have widened inequality and disparity issues. Every scholar concisely analyses the triggering factors of income inequality and disparity and suggest various means to overcome them to achieve SDGs. The authors have done appreciable jobs to gather inequality and disparity literature and made it easier for readers to find information on "international inequalities and disproportions towards SDGs" in one place.

There is no doubt that inequality and disparity have stratified the present-day society socially, culturally, economically, and environmentally and have unstable society with unseen social tensions for decades despite the long attempts made by the United Nations, governments, international and national agencies efforts to narrow the gaps between rich and poor. Editors (Profs. Bhandari and Shvindina Hanna) have taken the right initiatives to inform readers how these inequalities and disparities can be narrowed down and the how to create congenial working environment in this widely divided world. The editors' conclusion *"The invisible walls are everywhere and "my profit first" is the dominant approach of current development paradigms. To overcome the inequality problems the concept of "**Bashudaiva Kutumbakkam**" – The entire world is our home and all living beings are our relatives" and **Live and let other live** – the harmony within, community, nation and global" is needed"* is very inspiring to follow.

This book will be very useful to researchers, planners, policy makers, civil society, government, national and international development agencies and United Nations and its agencies. I would like to congratulate the authors and editors for this commendable job.

Keshav Bhattarai, PhD
Professor of Geography
School of Geoscience, Physics, and Safety
University of Central Missouri
Warrensburg, Missouri 64093, USA

This book "Reducing Inequalities Towards Sustainable Development Goals: Multilevel Approach" (edited by Profs. Medani P. Bhandari and Shvindina Hanna) provides a multidisciplinary approach to evaluate state of inequality in global, continental European Union, with country specific case study of Ukraine. The book shows poverty reduction in the developing world is still a major challenge and economic, social (gender, sex, race and invisible tradition driven) inequality has been increased despite of various intervention from concern stakeholders including United Nations; the environmental degradation continues to increase; climate change has created devastating impact already seen through melting artic and Himalaya, unpredictable rain pattern, landslides and drought; environmental health hazards impact particularly to the marginalized population of developing world are one way or another are also variables to increase inequality. Authors of this book incorporate underlying causes of inequality through theoretical lenses of development economics as well as social justice and dominance theories. Each chapter are unique in case presentation, however, reveal the factual evidences that inequalities is growing globally. Book provides the relevance of SDGs as a hope to reduce inequalities (social, economic, environmental); however, the results are donot show any rational impact in reducing socioeconomic strata in the society. Book is very interesting to read and useful to all concerned stakeholders, academia, governments, civil societies, governments, developmental agencies and UN agencies.

Tulsi Dharel MBA, PhD
Professor of Marketing and International Business The Business School
Centennial College, Toronto, Canada

In their book Reducing Inequalities Towards Sustainable Development Goals: Multilevel Approach, Professors Medani P. Bhandari and Shvindina Hanna broach new horizons related to the linkages between inequality and sustainable development. The authors look deep into the SDG Goal 10 "Reduced Inequalities", which lists several targets to improve the existing inequalities in income and wealth and offer deeper insights into some of the targets and their implications. This book provides an excellent analysis related to how increasing inequalities can hamper sustainable development efforts. For example, the current debate in the United States related to universal health care, access to education and living wages highlights how even in one of the richest and technologically most advanced countries, inequality can threaten social harmony and the well-being of its citizens impacting the goal of sustainable development. This book is an important and timely contribution towards understanding how income and wealth inequality can threaten the hard-earned gains of societies across the world and derail their ongoing work towards sustainable development. The data-rich chapters, references, and analysis contained in the book make the book a recommended reading for development professionals, policy makers and students interested in the nexus between inequality and development especially in developing countries. The authors have provided a useful book to help us understand an important topic of our times.

Dr. Ambika P. Adhikari – Former Country Representative of IUCN in Nepal, and former Research Professor, and Senior Sustainability Scientist at Arizona State University. Currently, he is a Principal Planner heading the Long-Range Planning Division at City of Tempe in Arizona, USA
12 July 2019

This book by *Medani P. Bhandari and Shvindina Hanna* basically aims to address intra and international inequities in the context of achieving sustainable development goals (SDGs) of the nation states. It raises some important issues pertaining to the concept of sustainable development and the methodological requirements and approaches for achieving SDGs. The authors of the book have attempted to succinctly describe the ubiquitous nature of inequities and its multidimensional facets. The book demonstrates how global inequities is increasing despite nation states' apparent commitment to achieving SDGs. Authors argue that all three pillars of sustainability namely social, economic and environmental have been plagued with misconceptions, challenges and faulty methodological approaches which stand in the way of achieving SDGs.

The book consists of 12 chapters representing conceptual, theoretical, methodological and strategic approaches in understanding and achieving SDGs of the nation states. The final chapter is devoted to the sustainable development strategies in the context of fourth industrial revolution. The book, by and large, is useful for the students and the professionals who wish to develop a better understanding of the concept of sustainable development, its challenges and the methodological approaches for achieving SDGs of the nation states.

Dr. Gopi Upreti
Retired Professor, Tribhuvan University, Nepal

While many nations across the globe are experiencing higher level of infrastructural development, use of modern communication technologies, access to advanced health facilities, improved education system, and increased employment opportunities in recent decades, gaps between and within the societies on socio-economic conditions, public health, quality education, and even the degree of exposures to natural disasters are widening day by day. Increasing gaps between haves and haves not is not only resulting on segregated societies in relation to the income levels, consumption patterns, and living conditions but also affecting heavily on natural resource utilization, conservation, and sustainable development. Increasing inequality is transmitting strong, often irrational, signals especially to the lower income and disadvantageous groups in terms of their production relationships, resource extraction, environmental conservation, and ecological integrity. In this context, this edited book "Reducing Inequalities Towards Sustainable Development Goals: Multilevel Approach" by Medani P. Bhandari and Shvindina Hanna is a very timely publication and serves as a highly valuable resource for planners, resource managers, academicians, students, researchers, politicians, and development specialists. This book gives new insights into the interconnected passion of United Nations to reduce inequality through sustainable development.

Durga D. Poudel, PhD
Professor and Coordinator, Environmental Sciences, MS, BS Director, Ag Auxiliary Units, Regent Professor in Applied Life Sciences South Louisiana Mid-Winter Fair/ BORSF, University of Louisiana Lafayette, Lafayette, Las Angelus, USA
July 14, 2019

Medani P. Bhandari and Shvindina Hanna edited book, "Reducing Inequalities Towards Sustainable Development Goals: Multilevel Approach" contributes to our ability to achieve sustainable development goals (SDGs) internationally. The book highlights methodological requirements and approaches. The authors have shared their experience here and emphasize the importance of all three pillars of sustainability: social, economic and environmental. This book is useful for students and professionals in obtaining a more informed view for facing common challenges in sustainable development.

Vincent G. Duffy, Associate Professor of Industrial Engineering and Agricultural and Biological Engineering, Purdue University, USA

This book edited by Prof. Medani P. Bhandari and Associate Prof. Shvindina Hanna titled "Reducing Inequalities Towards Sustainable Development Goals: Multilevel Approach" depicts a chronological and categorical approach to evaluate state of inequalities in global affairs such as continental European Union, with country specific case study of Ukraine. The book unveils poverty reduction in the developing world is still a major challenge. There is digital divide, economic divide and so on across the board. The book has artistically articulated these variances. One of the mind bugling questions that the book raised is that inequality has been increased despite of various intervention from concern stakeholders including United Nations; the environmental degradation continues to increase; climate change has created devastating impact already seen through melting artic and Himalaya, unpredictable rain pattern, landslides and drought; environmental health hazards impact particularly to the marginalized population of developing world are one way or another are also variables to increase inequality. The authors seemed to study the cause and effect of this chain using theoretical lenses of developmental economics as well as social justice. The book is so interesting that it can be useful reference book to all spectrum like academia, governments, civil societies, governments, developmental agencies including UN agencies.

Medani Adhikari MA, MSC
Professor of Economics at American Military University, Charles Town, WV
Founder and CEO, Equality Foundation, VA
Chief Executive Officer, Micro Logistics Inc. VA

The book entitled "Reducing Inequalities Towards Sustainable Development Goals: Multilevel Approach" edited by Prof. Medani P. Bhandari and Associate Professor Shvindina Hanna, is most probably the first book which tries to see the interlinkages of economic inequality and social and cultural invisible walls, at global, intercontinental – Europe and provides the case studies of Ukraine. It is found that writers have tried to carry out the fact through their scholar study that economic inequality is has been a major challenge throughout human history. However, United Nations has been trying to resolve this problem through its programs since its inception, the current programs and policies of sustainable development are new approached to minimize inequality globally and particularly in the developing world. We believe this study has stablished a new dynamism in utilization of socio-economic, developmental and political agendas. Governments, planners, educators and socio-economic interpreters can directly be benefited by this book.

Guna Raj Luitel,
Global President, International Network for Nepali Journalists (INNJA, USA)
Senior IT Specialist, US- Federal Government System, Washington DC, USA

Kaerbergen July 23th, 2019

To whom this may concern,

The monograph is devoted to a series of core aspects on the progress towards sustainable development. Experts, practitioners and educators are united in their thoughts about increasing inequalities and disparities between different regions and social groups. Considering the significance of sustainable development goals, this book offers a most attractive collection of academic works offered by authors who are acknowledged experts in their fields. Co-editor, Professor P. Medani Bhandari, an Associate Professor Sumy State University performed a unique job, in bringing together authors from different institutions. This allowed a process of developing the ideas, methods and strategies to overcome barriers towards a better common future.

This volume consists of chapters that cover the problems of an unbalanced social, environmental and economic development. I am very pleased to support my colleagues in their intentions to accelerate achieving the sustainable development goals by decreasing the unfairness and inequalities. This book will be helpful for policy-makers, researchers, decision makers and non-government agencies as it encourages identifying and overcoming the challenges in sustainable development.

Em. Prof. Dr. Luc Hens

Vlaamse Instelling voor Technologisch Onderzoek (VITO)

Boeretang 200

B2400 Mol, Belgium

Acknowledgements

First of all, I would like to thank my colleague, co-editor of this volume Associate Professor Shvindina Hanna and all members of her family (parents, Olexandr and Valentina, and best friend and supporter Arseniy Prokhasko) and Prof. Oleksandr Telizhenko, Prof. Volodymyr Boronos, Prof. Tetyana Vasilyeva, Prof. Oleg Balatskyi (her mentor 1999–2012) and thank Professor Duffy from Purdue University, who created an excellent environment by supporting her to devote herself in this book project. I should acknowledge that, she is the key manager of this project – making initial book plan to establishing contacts with authors – collecting manuscripts, several round editing (back and forth with authors) and other editorial tasks as needed. We most remember Daryna Leus (Boronos), who was one of important proponent, of this book, unfortunetly she is no more with us, we never forget her – we dedicate this book to her.

Equally, I would also like to thank to all authors who contributed in this volume – Professors of Sumy State University – Anna Buriak, Olena Chygryn, Iryna D'yakonova, Iryna Dehtyarova, Oleksandra Karintseva, Olexandra Karintseva, Mykola Kharchenko, Nadiya Kostyuchenko, Oleksandr Kubatko, Serhiy Lyeonov, Oleksii Lyulyov, Leonid Melnyk, Tetyana Pimonenko, Olena Shkarupa, Shvindina Hanna, Denys Smolennikov, Iryna Sotnyk, Mykola Sotnyk, Svitlana Tarasenko, Tetyana Vasilyeva, Olexander Zaitsev, Professor of Ukrainian-American Concordia University Zharova Liubov, and young researchers and PhD students Rodionova P., Obod O., Dyachenko A.V. and Artemenko O. Each of you have done excellent job, your scholarly work bridges the knowledge gap in the complex field – inequality with reference to sustainable development goals.

I would like to thank Prof. Douglas Capogrossi (my mentor, President of Akamai University, Hawaii), Prof. Anatoliy Vasylyev, (Rector of Sumy State University), Prof. Bishnu Paudel (Guru of all of us), Mr. Kedar Neupane (UNHCR-Geneva), Dr. Ambika Adhikari (mentor for environment conservation), Prof. Gopi Uprety (Tribhuvan University, Nepal), Professor Durga D. Paudel (University of Louisiana Lafayete), Prof. Tulsi Dharel, (Centennial

College, Toronto, Canada), Prof. Cecelia Green (Syracuse University), Prof. Keshav Bhattarai (Arizona State University), Prof. Vincent Duffy (Purdue University), Mr. Medini Adhikari (Equality Foundation), Mr. Guna Raj Luitel (INJA-Global), and Dr. Oleksander Romanovskiy (Rector of Ukrainian-American Concordia University, Kyiv, Ukraine) for their reviews, forwards and endorsements. Your togetherness with us adds value in this book project.

Especial thanks to Prof. Boronos Volodymyr, Head of the Department, of Finance and Entrepreneurship, I never forget your encouragements during my stay in Ukraine. You have given me new insights specially to dig on Marxist philosophy. I would also like to thank all my department colleagues Profs. (Dr.) Vasilyeva Tetyana, Kobushko Ihor, Basantsov Ihor, Hrytsenko Larysa, Dr. Salvin Paul, Aleksandrov Vadym, Abramchuk Maryna, Zaitsev Oleksandr, Zakharkin Oleksii, Zakharkinà Liudmyla, Illiashenko Kostiantyn, Illiashenko Tetiana, Humenna Yuliia, Kasianenko Tetiana, Kostel Mykola, Kotenko Natalia, Rubanov Pavlo, Pokhylko Svitlana, Plikus Iryna, Saltykova Hanna, Skliar Iryna, Shkodkina Yuliia, Tiutiunyk Inna, Zhukova Tetiana, Mohylnyi Viktor. You all are amazing, your knowledge sharing expertise is commendable. Thank you to you all.

Thank you Prof. Oleksandr M. Telizhenko, Head of the Department, Department of Management and all deputy heads Professors. Baistriuchenko Nataliia, Martynets Viktoriya, Zhulavskyi Arkadii, and Taraniuk Karina for your support to this book project. I would also like to thank all faculties and researchers at the department – Professors/Drs. Bondar Tetiana, Demikhov Oleksii, Galinska Yulia, Gryshchenko Iryna, Gordienko Vita, Degtyarenko Oleksandr, Dreval Olga, Yevdokymova Alona, Kyslyi Volodymyr, Kobushko Iana, Kolosok Svitlana, Kubatko Victoria, Lukianykhin Vadym, Mayboroda Tetyana, Matveieva Yuliia, Myroshnychenko Iuliia, Mishenina Galyna, Onishchenko Marharyta, Opanasiuk Yuliia, Pavlenko Olena, Panchenko Olga, Petrushenko Mykola, Rybalchenko Svitlana, Smolennikov Denys, Tyhenko Volodymyr, Shevchenko Hanna, Vakulenko Ihor and Valenkevych Larysa. I would like to thank to all department specialists – Koval Yuliia, Litsman Viktoriia and Tokaryeva Larysa, for your support in this project.

I would like to thank to my students Viktoria Makerskaya and Yaroslav Reshetnyak for updating about the inequality situation in Ukraine. I acknowledge Konstantin Moskalenko for constant reminders about the rights of disable people and their problems, how they are treated in the society and even at home environment. Special thanks to Sergey Mayboroda, HDR Market (c), for donation of the cover page picture for this book.

I would also like to thank to my wife Prajita Bhandari for her encouragements and for giving insightful information on how women are victims of inequality-at home to work environment. I would also like thank to our son Prameya, daughter Manaslu and daughter in law Kelsey for helping me to find relevant resources in the field of inequality, social division, stratifications and gender issues. I would also like to thank to my mother Hema Devi, brothers Krishna, Hari, sisters Kali, Bhakti, Radha, Bindu, Sita and their families for encouraging me by providing peaceful environment. I would also like to thank Guru Jee and Guru Ma of Saidam Canada, all members of Sai Group of Virginia and my friends Rajan Adhikari, Govinda Luitel, Tirtha Koirala, Sanjay Mishra, Bijay Kattel, Dhir Prasad Bhandari and all of my facebook and LinkedIn's friends, who have been always encouraging to us to give back to the society through knowledge sharing. Thank you to Mark de Jongh, Rajeev Prasad and Junko Nakajima of River Publishers, for encouraging and empowering us to complete this book project on time. Thank you to you all, who have given their inputs for this book project directly or indirectly.

Thank you,

Medani P. Bhandari

List of Contributors

Anna Buriak, *Department of Finance, Banking and Insurance, Sumy State University, Ukraine*; *E-mail: ann.v.buriak@gmail.com*

Anna Dyachenko, *Department of Economics and Business-Administration, Sumy State University, Ukraine*

Denys Smolennikov, *Department of Management, Sumy State University, Ukraine*; *E-mail: dos@management.sumdu.edu.ua*

Iryna Dehtyarova, *Department of Economics, Entrepreneurship and Business Administration, Sumy State University, Ukraine*

Iryna D'yakonova, *Department of International Economic Relations, Sumy State University, Ukraine*; *E-mail: i.diakonova@uabs.sumdu.edu.ua*

Iryna Sotnyk, *Department of Economics, Entrepreneurship and Business Administration, Sumy State University, Ukraine*
E-mail: insotnik@gmail.com

Leonid Melnyk, *Department of Economics, Entrepreneurship and Business Administration, Sumy State University, Ukraine*
Institute for Development Economics of Ministry for Education and Science, National Academy of Science, Sumy State University, Ukraine
E-mail: melnyksumy@gmail.com

Medani P. Bhandari, *Akamai University, USA*
Finance and Entrepreneurship Department, Sumy State University, Ukraine
E-mail: medani.bhandari@gmail.com

Mykola Kharchenko, *Department of Economics, Entrepreneurship and Business Administration, Sumy State University, Ukraine*

Mykola Sotnyk, *Research Institute of Energy Efficient Technologies, Sumy State University, Ukraine*

Nadiya Kostyuchenko, *Department of International Economic Relations, Sumy State University, Ukraine*

Oleksandr Artemenko, *Department of Finance, Banking and Insurance, Sumy State University, Ukraine*

Oleksandr Kubatko, *Department of Economics, Entrepreneurship and Business Administration, Sumy State University, Ukraine*

Oleksandra Karintseva, *Department of Economics, Entrepreneurship and Business Administration, Sumy State University, Ukraine*

Oleksii Lyulyov, *Marketing Department, Sumy State University, Ukraine*

Olena Chygryn, *Marketing Department, Sumy State University, Ukraine*

Olena Obod, *Department of International Economic Relations, Sumy State University, Ukraine*

Olena Shkarupa, *Department of Economics, Entrepreneurship and Business Administration, Sumy State University, Ukraine*; *E-mail: elenashkarupa@econ.sumdu.edu.ua*

Olexander Zaitsev, *Finance and Entrepreneurship Department, Sumy State University, Ukraine*

Polina Rodionova, *Department of International Economic Relations, Sumy State University, Ukraine*

Serhiy Lyeonov, *Economic Cybernetics Department, Sumy State University, Ukraine*

Shvindina Hanna, *Department of Management, Sumy State University, Ukraine*
Purdue University, USA; *E-mail: shvindina.hannah@gmail.com*

Svitlana Tarasenko, *Department of International Economics, Sumy State University, Ukraine*; *E-mail: svitlana_tarasenko@ukr.net*

Tetyana Pimonenko, *Marketing Department, Sumy State University, Ukraine*; *E-mail: tetyana.pimonenko@gmail.com*

Tetyana Vasilyeva, *Finance and Entrepreneurship Department, Sumy State University, Ukraine*

Zharova Liubov, *Wyższa Szkoła Ekonomiczno-Humanistyczna/ The University of Economics and Humanities, Bielsko-Biala, Poland*; *E-mail: zharova.l@gmail.com*

List of Figures

List of Tables

1

Introduction: The Problems and Consequences of Sustainable Development Goals

Medani P. Bhandari[1,2,*] and Shvindina Hanna[3,4]

[1]Akamai University, USA
[2]Finance and Entrepreneurship Department, Sumy State University, Ukraine
[3]Department of Management, Sumy State University, Ukraine
[4]Purdue University, USA
E-mail: medani.bhandari@gmail.com
*Corresponding Author

1.1 Introduction

This edited book, "Reducing Inequalities Towards Sustainable Development Goals: Multilevel Approach" consists of chapters written by scholars and researchers mainly from USA and Ukraine.

Most of the authors have extensive experience from USA, Europe, and Post-Soviet nations, and all contributing authors have rich and extended academic and research experiences in their respected fields they are dealing in this book. The chapters outlined at the end of this chapter and authors brief biography is included at the end of this book.

Inequality is one of the major divisive factors of human civilization and the evolutionary history of development. Inequality can be seen in every sphere of social, political, and economical structure of community, national and international level. Such unequal strata can be grounded on innumerable factors – such as social, cultural, political, geographical or due to environmental (anthropogenic or natural) catastrophe. "To speak of a social inequality is to describe some valued attribute which can be distributed across the relevant units of a society in different quantities, where 'inequality' therefore implies that different units possess different amounts of this attribute. The units can be

1

individuals, social groups, communities, nations; the attributes include such things as income, wealth, status, knowledge, and power" (Wright 1994:21, as in Soares et al., 2014:1). Inequality is multidimensional and can be seen in complex forms – including assets, access to basic services, infrastructure, and knowledge, race, gender, ethnic and geographic dimensions (Soares et al., 2014). The concept of inequality also has human attitudinal aspect (happiness) (Brockmann and Delhey, 2010).

> *"a house may be large or small; as long as the neighboring houses are likewise small, it satisfies all social requirement for a residence. But let them arise next to the little house a palace, and the little house shrinks to a hut. The little house now makes it clear that its inmate has no social position at all to maintain, or but a very insignificant one; and however high it may shoot up in the course of civilization, if the neighboring palace rises in equal or even in greater measure, the occupant of the relatively little house will always find himself more uncomfortable, more dissatisfied, more cramped within his four walls"*
>
> (Karl Marx quoted as cited in Freistein and Mahlert, 2015:2–3)

This is very true; however, the disparity has different dimensions between haves and haves not. The shadow of haves plays critical role to raise the gap within societal strata, which amplifies to national and international level.

For example [Currently in the United States, the gap between the top 1% of income earners and the bottom 50% of income earners is back to levels not seen since the 1920's (Forbes, 2019). This gap can partly be attributed to economic conditions that financially favor corporations over workers. This division between the ultra-wealthy and common man is not a new phenomenon. This notion of fact applies to the global context as well, whereas the gap of rich and poor is increasing]. The World Inequality Report 2018 shows the similar trend. The report states that "since 1980, income inequality has increased rapidly in North America and Asia, has grown moderately in Europe, and has stabilized at extremely high levels in the Middle east, sub-Saharan Africa, and Brazil. The poorest half of the global population has seen its income grow significantly thanks to high growth in Asia (particularly in China and India). Perhaps the most striking finding of this report, however, is that, at the global level, the top 0.1% income group has captured as much of the world's growth since 1980 as the bottom half of the adult population. Conversely, income growth has been sluggish or even nil for the population between the global bottom 50% and top 1%. This includes North American and European lower- and middle-income groups.

The diversity of trends observed in the report suggest that global dynamics are shaped by a variety of national institutional and political context. There is no inevitability behind the rise of income inequality" (World Inequality Report, 2018:278). The following facts and figures provide a clear picture of inequality directions.

> *In recent decades, income inequality has increased in nearly all countries, but at different speeds, suggesting that institutions and policies matter in shaping inequality. Since 1980, income inequality has increased rapidly in north America, China, India, and Russia. Inequality has grown moderately in Europe (Figure 1.1). from a broad historical perspective, this increase in inequality mars the end of a postwar egalitarian regime which too deferent forms in these regions.*
>
> (World Inequality Report 2018:9–10)

> *There are exceptions to the general pattern. In the middle east, sub-Saharan Africa, and Brazil, income inequality has remained relatively stable, at extremely high levels Figure 1.2. Having never gone through the postwar egalitarian regime, these regions set the world inequality frontier. The diversity of trends observed across countries since 1980 shows that income inequality dynamics are shaped by a variety of national, institutional and political contexts.*
>
> (World Inequality Report 2018:10–11)

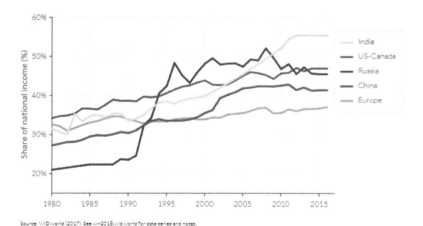

Source: WID.world (2017). See wir2018.wid.world for data series and notes.
In 2016, 47% of national income was received by the top 10% in US-Canada, compared to 34% in 1980.

Figure 1.1 Top 10% income shares across the world, 1980–2016: rising inequality almost everywhere, but at different speeds.

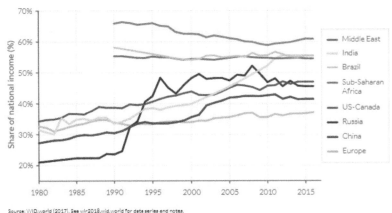

Source: WID.world (2017). See wir2018.wid.world for data series and notes.
In 2016, 55% of national income was received by the Top 10% earners in India, against 31% in 1980.

Figure 1.2   Top 10% income shares across the world, 1980–2016: Is world inequality moving towards the high-inequality frontier?

What is the future of global income inequality?

The future of global income inequality is likely to be shaped by both convergence forces (rapid growth in emerging countries) and divergence forces (rising inequality within countries). No one knows which of these forces will dominate and whether these evolutions are sustainable.

However, our benchmark projections show that if within-country inequality continues to rise as it has since 1980, then global income inequality will rise steeply, even under fairly optimistic assumptions regarding growth in emerging countries. The global top 1% income share could increase from nearly 20% today to more than 24% in 2050, while the global bottom 50% share would fall from 10% to less than 9%.

If all countries were to follow the high inequality growth trajectory followed by the United States since 1980, the global top 1% income share would rise even more, to around 28% by 2050. This rise would largely be made at the expense of the global bottom 50%, whose income share would fall to 6%.

Conversely, if all countries were to follow the relatively low inequality growth trajectory followed by Europe since 1980, the global top 1% income share would decrease to 19% by 2050, while the bottom 50% income share would increase to 13%.

> *Differences between high and low inequality growth trajectories within countries have an enormous impact on incomes of the bottom half of the global population. Under the US-style, high inequality growth scenario, the bottom half of the world population earns €4 500 per adult per year in 2050, versus €9 100 in the EU-style, low inequality growth scenario (for a given global average income per adult of €35 500 in 2050 in both scenarios).*
>
> (World Inequality Report 2018:250)

In addition to convergence forces and divergence forces inequality scenarios, there are also many historical contexts to create division within society, and beyond the countries (colonialism, wars, the racial supremacy concepts and other historical events are the key players to create and increase inequality). Historical condition constitutes the framework within which this creation takes place, each generation carries with it the entire history of the human species (MEW, 1953:115). The human being embodies history as a tradition and applies it to contemporary reality (Guidetti, 2014).

There are various approaches to inequality measurement (Guidetti, 2014). Theoretically, both qualitative and quantitative approaches can be applied; however, broadly quantitative approach is dominant in the economic domain. Most practised approaches (within the quantitative domain) are Human capital; The effect of skill-biased technical change; Internationalization of production; Labor market institutions; Role of the Welfare state; Inequality (The basic idea is that inequality can be conceived as an autoregressive process, where the degree of past inequality affects present inequality); Models of capitalism and institutional complementarities; The governance of firms (Guidetti, 2014). One of the most used measures of inequality is the analysis of Gini coefficient. An example of Gini coefficient to evaluate inequality:

Using quantitative approach, this book covers few social aspects of inequality (in terms of gender and labor distribution) and mostly elaborates on economic facets of inequality relation to Sustainable Development Goals.

> *"We must all realize that inequality reduction does not occur by decree; neither does it automatically arise through economic growth, nor through policies that equalize incomes downward via blind taxing and spending. Inequality reduction involves a collaborative effort that must motivate all concerned parties, one that constitutes a genuine political and social innovation, and one that often runs counter to prevailing political and economic forces"*
>
> (Genevey et al., 2013:6)

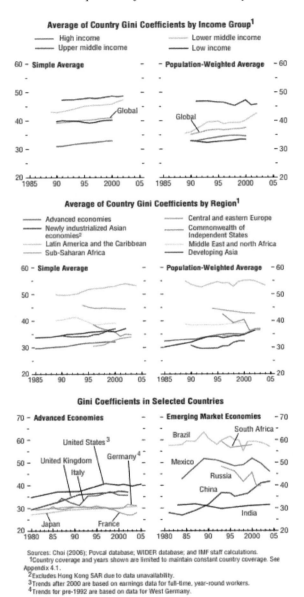

Figure 1.3   Cross-country trends in inequality (Gini coefficient).

As shown in the figures self-explanatory *(Inequality has risen in developing Asia, central and eastern Europe, the NIEs, and the advanced economies, while falling in the Commonwealth of Independent States and, to a lesser extent, in sub-Saharan Africa).*
Source: World Economic Outlook, 2007:36.

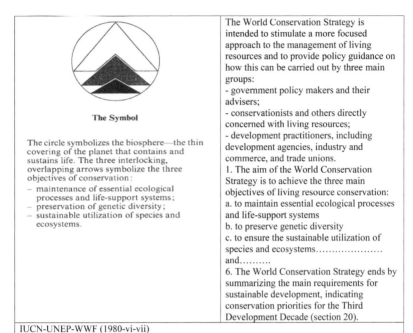

The World Conservation Strategy is intended to stimulate a more focused approach to the management of living resources and to provide policy guidance on how this can be carried out by three main groups:
- government policy makers and their advisers;
- conservationists and others directly concerned with living resources;
- development practitioners, including development agencies, industry and commerce, and trade unions.

1. The aim of the World Conservation Strategy is to achieve the three main objectives of living resource conservation:
a. to maintain essential ecological processes and life-support systems
b. to preserve genetic diversity
c. to ensure the sustainable utilization of species and ecosystems.....................
and..........
6. The World Conservation Strategy ends by summarizing the main requirements for sustainable development, indicating conservation priorities for the Third Development Decade (section 20).

The Symbol

The circle symbolizes the biosphere—the thin covering of the planet that contains and sustains life. The three interlocking, overlapping arrows symbolize the three objectives of conservation:
-- maintenance of essential ecological processes and life-support systems;
-- preservation of genetic diversity;
-- sustainable utilization of species and ecosystems.

IUCN-UNEP-WWF (1980-vi-vii)
https://portals.iucn.org/library/sites/library/files/documents/WCS-004.pdf

Figure 1.4 Sustainability symbol-by IUCN.

Sustainable development is a complex issue which interchangeably in use with Sustainability. Various authors have used the term in their own ways and change, weather variation (Corral-Verdugo et al., 2009; Betsy, 2010; SDSN, 2014; Boucher, 2015; United Nations, 2015; WHO, 2015; Mitchell and Walinga, 2017; Tahvilzadeh et al., 2017; Bhandari, 2018). The term sustainability discourse stands to maintain the equilibrium between nature and society and fulfill the societal demands (which could be environmental, economic and social). Sustainability scholarship is to search the know-how of how development can be maintained without hampering the natural ecosystem and how the global major problems i.e. Environmental problems, socioeconomic problems – poverty, hunger, health can be solved or at least minimize.

As such development of sustainability concept can be seen from IUCN, UNEP, WWF document "World Conservation Strategy: Living Resource Conservation for Sustainable Development" in 1980.

The world conservation strategy the symbol and the text clearly emphasize the importance to sustainability, which also paved the foundation of

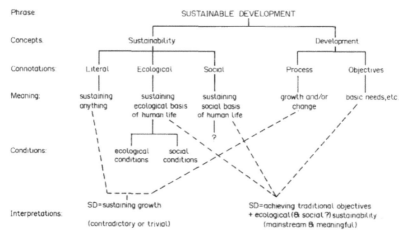

Figure 1.5   The semantics of sustainable development (as in Lele 1991:608).

Source: Lele 1991:608 (the existence of the ecological conditions necessary to support human life at a specified level of well-being through future generations, what I call ecological sustainability (Lele 1991:609). *"The current state of scientific knowledge (particularly insights obtained in the last few decades) about natural and social phenomena and their interactions leads inexorably to the conclusion that anyone driven by either long-term self-interest, or concern for poverty, or concern for intergenerational equity should be willing to support the operational objectives of SD"* (Lele 1991:612).

sustainability discourses. However, one can trace the originality of discourses when people began to think about the limitation of natural resources and interrelated harmonies relationships between human and nature. In this respect, we can see the modern environment conservation history and the efforts to conserve them particularly in terms of environmental problems – climate change. Sustainable development covers numbers of scenario of development economics.

Lele (1991) nicely presents the complexity of sustainable development (SD) from concepts to the implementing phase, which reveals the fact that SD depends on many aspects and can be implement with the application of multidimensional approaches. Table 1.1 gives general scenario of SD coverage (self-explanatory).

Sustainable development has just five decades of origin; however, it is already a dominant development paradigm. To take this concept at global level UN has playing instrumental role, through resolutions, development agendas, and commitments. Other international government and nongovernmental organizations and the UN member nations are using the term as remedy for all social, environmental and economic problem world has been facing. UN and member nations have been emphasizing the idea

Table 1.1   Coverage – sustainable development

Social	Environmental
Education	Freshwater/groundwater
Employment	Agriculture/secure food supply
Health/water supply/sanitation	Urban
Housing	Coastal Zone
Welfare and quality of life	Marine environment/coral reef protection
Cultural heritage	Fisheries
Poverty/Income distribution	Biodiversity/biotechnology
Crime	Sustainable forest management
Population	Air pollution and ozone depletion
Social and ethical values	Global climate change/sea level rise
Role of women	Sustainable use of natural resources
Access to land and resources	Sustainable tourism
Community structure	Restricted carrying capacity
Equity/social exclusion	Land use change
Economic	**Institutional**
Economic dependency/Indebtedness/ODA	Integrated decision-making
Energy	Capacity building
Consumption and production patterns	Science and technology
Waste management	Public awareness and information
Transportation	International conventions and cooperation
Mining	Governance/role of civic society
Economic structure and development	Institutional and legislative frameworks
Trade	Disaster preparedness
Productivity	Public participation

Source: UN-DSD, 2007; Bhandari, 2017.

in way that, this era can be considered as the regime of sustainability. As such the term is in use as a treatment of all problems. The definition of sustainable development became so lucid, vague, liquified and flexible and a solution. Additionally, the educational institutions are introducing the concept from the school to the highest level of education. UN is the main producer; however, Educational institutions are knowledge factories, seller and the knowledge seekers the graduate are consumer, reproducer and disseminator of this MANTRA. The effectiveness has to be tested. The United Nations agenda 21, millennium development goals and now sustainable development goals can be considered as examples of test mechanism and the UN member governments who has adopted these goals in their national plans and policies are considered as laboratory. The officials who involved in such policy formulation and the other personnel who are trying to implement are the lab technicians. These all involved personnel, wherever they work including

UN, Development agencies, International governmental or nongovernmental organizations, government agencies, private sectors, etc. are the product of educational institutions. Therefore, in creating sustainable development discourse, epistemology and regime; the educational institutions have playing a vital role. The sustainability paradigm has been shifting as the horizon of knowledge expanded. The sustainable development is no more just a concept but became an accepted policy, norms, and value of the development agendas. Linking with the institutional goals of sustainability, still the implementing agencies "the governments" have to prepare the institutional architecture (Bhandari, 2018).

The diverse characteristics of sustainable development has been in debate; however, it is has emerged as maturing domain of development with future hope. The United Nations and it's agencies have been using sustainable development as catchphrase in most of their development and nature related agendas (WCED, 1987; UNEP, 2012; UN, 2015; UNFCCC, 2016). Similarly, international development agencies, international nongovernmental organizations, government agencies, nongovernmental organizations embraced sustainable development as new paradigm of development (Klarin, 2018). The higher educational institutions are also playing phenomenon roles on theorizing the sustainability domain as well as evaluating practicality of this new development paradigm (Vaughter et al., 2013; Caeiro et al., 2013; Findler et al., 2019; AASHE, 2018; Bedenlier et al., 2017; UN, 2018). The concept is still in formalizing process. Table 1.2 gives the order of sustainable development – formalization process.

Table 1.2 reveals two diverse trajectories of sustainable development. First, it shows the maturation process and secondly, its changing and broadening meaning particularly in fostering as new scholarship which combines three basic coverage of sustainable development, social, economic and environmental discourse. This indicates that sustainable development is diverse, complex and complicated term, which is still in the process of validation. The United Nations and its member nations, the development agencies, international governmental and nongovernmental organizations are treating sustainable development as a remedy of the major problems the world is facing though agenda 21, MDGs and lately SDGs "sustainable future for all" and SDGs addresses the challenges we face, including those related to poverty, inequality, climate, environmental degradation, prosperity, and peace and justice (UN, 2015). *"The Sustainable Development Goals are the blueprint to achieve a better and more sustainable future for all. They address the global challenges we face, including those related to poverty, inequality,*

Table 1.2 The chronological overview of the meaning of sustainable development in the period 1980–2018

Authors/Publication and Year	Meaning and Understanding of Sustainable Development
IUCN, 1980	World Conservation Strategy
WCED, 1987	Sustainable development is a development that meets the needs of the present without compromising the ability of future generations to meet their own needs.
Pearce et al., 1989	Sustainable development implies a conceptual socio-economic system which ensures the sustainability of goals in the form of real income achievement and improvement of educational standards, health care and the overall quality of life
Harwood, 1990	Sustainable development is unlimited developing system, where development is focused on achieving greater benefits for humans and more efficient resource use in balance with the environment required for all humans and all other species.
IUCN, UNDP & WWF, 1991	Sustainable development is a process of improving the quality of human life within the framework of carrying capacity of the sustainable ecosystems.
Lele, 1991	Sustainable development is a process of targeted changes that can be repeated forever.
Meadows, 1998	Sustainable development is a social construction derived from the long-term evolution of a highly complex system – human population and economic development integrated into ecosystems and biochemical processes of the Earth.
PAP/RAC, 1999	Sustainable development is development given by the carrying capacity of an ecosystem.
Vander-Merwe & Van-der-Merwe, 1999	Sustainable development is a program that changes the economic development process to ensure the basic quality of life, protecting valuable ecosystems and other communities at the same time.
Beck & Wilms, 2004	Sustainable development is a powerful global contradiction to the contemporary western culture and lifestyle.
Vare & Scott, 2007	Sustainable development is a process of changes, where resources are raised, the direction of investments is determined, the development of technology is focused, and the work of different institutions is harmonized, thus the potential for achieving human needs and desires is increased as well.

(Continued)

Table 1.2   Continued

Authors/Publication and Year	Meaning and Understanding of Sustainable Development
Sterling, 2010	Sustainable development is a reconciliation of the economy and the environment on a new path of development that will enable the long-term sustainable development of humankind.
Marin et al., 2012	Sustainable development gives a possibility of time unlimited interaction between society, ecosystems and other living systems without impoverishing the key resources.
Duran et al., 2015	Sustainable development is a development that protects the environment, because a sustainable environment enables sustainable development.
Bhandari, 2018	Sustainable development is a fundamental basis of development practice and way of thinking ahead.

Source: Klarin, 2018:77.

Table 1.3   SDG goals

1. No Poverty	10. **Reduced Inequality**
2. Zero Hunger	11. Sustainable Cities and Communities
3. Good Health and Well-being	12. Responsible Consumption and Production
4. Quality Education	13. Climate Action
5. Gender Equality	14. Life Below Water
6. Clean Water and Sanitation	15. Life on Land
7. Affordable and Clean Energy	16. Peace and Justice Strong Institutions
8. Decent Work and Economic Growth	17. Partnerships to Achieve the Goal
9. Industry, Innovation and Infrastructure	

Source: United Nations, 2015.

climate, environmental degradation, prosperity, and peace and justice. The Goals interconnect and in order to leave no one behind, it is important that we achieve each Goal and target by 2030" (United Nations, 2015).

1.2 Inequality a Major Challenge

United Nations (2015) clearly acknowledges that inequality is one of major challenge of contemporary world. As an evidence "Out of the 17 goals, eleven address forms of inequality, in terms of equality, equity and/or inclusion (Goals 1, 3, 4, 5, 6, 7, 8, 9, 10, 11, 16 and 17), and one goal (Goal 10) explicitly proposes to reduce various forms of inequalities" (Freistein and Mahlert, 2015:7) (Table 1.3).

Goal 10 has seven targets.

1. **10.1** – By 2030 progressively achieve and sustain income growth of the bottom 40% of the population at a rate higher than the national average.
2. **10.2** – By 2030 empower and promote the social, economic and political inclusion of all irrespective of age, sex, disability, race, ethnicity, origin, religion or economic or other status.
3. **10.3** – Ensure equal opportunity and reduce inequalities of outcome, including through eliminating discriminatory laws, policies and practices and promoting appropriate legislation, policies and actions in this regard.
4. **10.4** – Adopt policies especially fiscal, wage, and social protection policies and progressively achieve greater equality.
5. **10.5** – Improve regulation and monitoring of global financial markets and institutions and strengthen implementation of such regulations.
6. **10.6** – Ensure enhanced representation and voice of developing countries in decision making in global international economic and financial institutions in order to deliver more effective, credible, accountable and legitimate institutions.
7. **10.7** – Facilitate orderly, safe, regular and responsible migration and mobility of people, including through implementation of planned and well-managed migration policies (United Nations, 2015).

Importantly, these targets clearly identify the major indicators of inequality (measurable or unmeasurable) – age, sex, disability, race, ethnicity, origin, religion or economic or other status; inequalities of outcome, including through eliminating discriminatory laws, policies and practices; labor wage, and developing – developed countries variation.

1.3 Sustainable Development Challenges

Global sustainable development challenges have been highlighted through various publication as well as the United Nations itself and its agencies (Daly, 1991, 1996; Abdallah et al., 2009; Bhandari, 2017, 2018; DESA, 2013; UNEP, 2010; UNESCO, 2013; United Nations, 2015; United Nations General Assembly, 1987; Berg and Ostry, 2011; Galbraith, 2012; Jackson, 2009).

"The world is faced with challenges in all three dimensions of sustainable development – economic, social and environmental. More than 1 billion people are still living in extreme poverty, and income inequality within and among many countries has been rising; at the same time, unsustainable consumption and production patterns have resulted in huge economic and social costs and may endanger life on the planet. Achieving sustainable development will require global actions to deliver on the legitimate aspiration towards further economic and social progress, requiring growth and employment, and at the same time strengthening environmental protection" (DESA, 2013:iii).

These global actions are possible to implement under institutions transformation and educational initiatives. It is obvious that customer individual behavior is a core of market changes, and market changes together constitute the global trends. The search for a balance between social, economic and environmental pillars is possible under conditions of developing educational programs on green economics and interdisciplinary research. As far as environmental engineering, climate mitigation and environmental economics are related, many social initiatives exist disconnected at different local communities. For example, the accumulation of rainwater at special reservoirs seems to be "green initiative" and accepted by the local community but leads to degradation of the nearest soil and water bodies. The consequences of these decisions will be obvious a time later. Education in a sphere of Sustainable Development is important, as well as education in a sphere of green decisions for local communities in correspondence to global goals.

However, this seems very ambitious presumption since the challenges are increasing in alarming scale. There is a need of empowering the societies to cope with the challenges, by which they will be enable write own pathway and future directions.

1.4 Book Chapters Arrangement

As noted above, the first chapter provides the theoretical background of sustainable development and its goal set, SDG challenges and gives general introduction of the book. Chapter 2, highlights the problems of inequalities, disproportions and disparities in the process of SDGs implementation and also proposes the win-win strategy for the regions and territories to SDGs implementation. The overview of the existing concepts allowed to reveal the disparities in economic growth and contradictories in using terms related to equality, stability, climate change, poverty. The chapter presents the shift towards identifying new challenges in a sphere of economic development. Chapter 3, explores more on inequalities in relation to economic development. Why equality matters. The concept of "re-thinking" of modern management approaches. Generalizing of ways of overcoming common barriers of economic development. Chapter 4 is about green investment: definition, features, mechanism to implement. Provides a comparison analyses of funding and stimulating green projects in EU and Ukraine, and perspectives to provide and spread green investments in Ukraine.

Chapter 5 provides general concepts of public institutions and good governance in relation to trust building and also defines the relevant factors to reach SDGs considering social conditions and economic abilities of different countries in order to increase the efficiency of expenditures through enhancing of institutions. Chapter 6 provides a historical overview of labor wages comparison, gender inequality and doubts on achieving SDGs in relation to variation in labor market. It shows how theory of labor value is the development of the economy towards the civil economy. Indicators and measures of human labor costs are taken as main framework to analysis the inequality problem. Difference and interaction of monetary and labor indicators became a center of discussion in this chapter which led to presentation of a new concept of a sustainable economy based on labor value. Chapter 7 emphasizes the venture capital as a component of finance market. The institutional models in a sphere of venture capital development are generalized taking into account EU experience and global tendencies. The best practices of venture companies are performed to investigate the possibilities of implementation of their patterns into Ukraine reality. Chapter 8 unveils the experience of implementation of the innovations for the SD on the base of EU experience. It also focuses on the problems of costs of economic restructuring of national economy.

Chapter 9 provides a general account of economic, social, political, and environmental benefits of renewables (RE) draw the attention of policymakers to the issue of deploying green power capacities. It examines the preconditions, trends and mechanisms of RE development worldwide and in Ukraine, analyzes the issues of alternative energy sector development for domestic households as the youngest RE market participant. Chapter 10 presents the importance of social responsibility for sustainable development goals. This chapter shows the connection between different SDGs and the links with social responsibility. Chapter 11 analyzes motivational tools on decision making process. The motivation is an impulse to act which is determined certain motive or group of the motives; it is the process of choosing between different possible actions that define the behavior of the subjects focused. The system software of sustainable development supposes using of motivational toolset, enabling the consumption of environmental goods and services, and the development, implementation environmental technologies.

Chapter 12 discusses about sustainable development strategies in conditions of the 4th industrial revolution: the EU experience. It reflects information, economic and technological transformations for ensuring sustainable development. It shows the impact areas, such as production, consumption and interface where environmentally friendly decisions necessary for responding to current socio-economic and environmental challenges in the EU and Ukraine. It highlights problems, methods incentive instruments for ensuring sustainable development management and greening economy.

This book encourages to rethink economic development through the lens of growing inequalities and disparities. These disparities became obvious with sharing information about global changes and distribution of environmental, economic and social risks among the recipients. The economic growth per se is disproportional and the efforts of scholars, practitioners and policymakers should be directed to find the balance between three pillars of sustainable development towards achieving SDGs, in particular, SDG 10 (Reducing inequalities). The book presents new insights for evaluating the progress on SDGs, sets new economic, social and environmental targets. The main challenges and focus of every chapter are different and yet in combination they give an integrated understanding of the phenomenon of sustainable development and its diverse aspects. This book will be useful for policymakers, social and environmental activists, agencies, educators and practitioners in the sphere of environmental economics. The methodology of the research can be replicated and taken forward by future researchers in the field.

References

AASHE (2018). Stars Technical Manual. (The Association for the Advancement of Sustainability in Higher Education). Version 2.1. Available online: https://stars.aashe.org/pages/about/technical-manual.html.

Adomßent, M., Fischer, D., Godemann, J., Herzig, C., Otte, I., Rieckmann, M. and Timm, J. (2014). Emerging Areas in Research on Higher Education for Sustainable Development – Management Education, Sustainable Consumption and Perspectives from Central and Eastern Europe, 62, 1–7. http://dx.doi.org/10.1016/j.jclepro.2013.09.045.

Adomßent, M., Godemann, J., Leicht, A. and Busch, A. (Eds.), (2006). Higher Education for Sustainability: New Challenges from a Global Perspective. VAS – Verlag für Akademische Schriften, Frankfurt am Main.

Alghamdi, N., den Heijer, A. and de Jonge, H. (2017). Assessment Tools' Indicators for Sustainability in Universities: An Analytical Overview. Int. J. Sustain. High. Educ., 18, 84–115.

Arima, A. (2009). A plea for more education for sustainable development. Sustain. Sci., 4(1), 3–5.

Barbier, E. B. (1987). The concept of sustainable economic development, Environmental Conservation, 14(2), 101–110.

Barker, Chris (2004). The SAGE Dictionary of Cultural Studies, SAGE Publications, London/Thousand Oaks/New Delhi. https://zodml.org/ sites/default/files/%5BDr_Chris_Barker%5D_The_SAGE_Dictionary_ of_Cultural_0.pdf.

Barnosky et al. (2012). Approaching a state shift in Earth's biosphere. Nature, 486(7401), 52–58.

Barth, M. and Godemann, J. (2006). Study program sustainability – a way to impart competencies for handling sustainability? In: Adomßent, M., Godemann, J., Leicht, A., Busch, A. (eds.) Higher education for sustainability: new challenges from a global perspective. VAS, Frankfurt, pp. 198–207.

Barth, M. and Michelsen, G. (2013). Sustain. Sci., 8:103. https://doi.org/10 .1007/s11625-012-0181-5 https://link.springer.com/article/10.1007/s1 1625-012-0181-5#.

Barth, M., Rieckmann, M. and Sanusi, Z. A. (Eds.), (2011). Higher Education for Sustainable Development. Looking Back and Moving Forward. VSA – Verlag für Akademische Schriften, Bad Homburg.

Baumert, Kevin A., Timothy Herzog and Jonathan Pershing (2005). Navigating the Numbers: Greenhouse Gas Data and International Climate Policy. Washington, D.C.: World Resources Institute.

Beall, Jo, Basudeb Guha-Khasnobis and Ravi Kanbur (Eds.), (2012). Urbanization and Development in Asia. Multidimensional Perspectives. New York: Oxford University Press.

Berg, Andrew and Jonathan Ostry (2011). Inequality and unsustainable growth: two sides of the same coin? IMF Staff Discussion Note. SDN/11/08. Washington, D.C.: International Monetary Fund. 8 April.

Bertelsmann Stiftung and Sustainable Development Solutions Network (2018). SDG Index and Dashboards Report 2018 – Global Responsibilities, Implementing the Goals, G20 and Large Countries Edition. www.pica-publishing.com, http://www.sdgindex.org/assets/files/2018/ 00%20SDGS%202018%20G20%20EDITION%20WEB%20V7%2018 0718.pdf.

Berzosa, A., Bernaldo, M. O. and Fernández-Sanchez, G. (2017). Sustainability Assessment Tools for Higher Education: An Empirical Comparative Analysis. J. Clean. Prod., 161, 812–820.

Bhandari, Medani, P. (2017). Climate change science: a historical outline. Adv. Agr. Environ. Sci., 1(1), 1–8: 00002. http://ologyjournals.com/aa eoa/aaeoa_00002.pdf.

Bhandari, Medani, P. (2018). Green Web-II: Standards and Perspectives from the IUCN, Published, sold and distributed by: River Publishers, Denmark/the Netherlands. ISBN: 978-87-70220-12-5 (Hardback) 978-87-70220-11-8 (eBook).

Bhattacharyya, Subhes, C. (2012). Energy access programs and sustainable development: A critical review and analysis, Energy for Sustainable Development, 16(3), 260–271.

Brockmann, Hilke and Jan Delhey (2010). The Dynamics of Happiness. Social Indicators Research, 97(1), 387–405.

Caeiro, S., Jabbour, C. and Leal Filho, W. (2013). Sustainability Assessment Tools in Higher Education Institutions Mapping Trends and Good Practices around the World; Springer: Cham, Gemany, p. 432.

Callanan, Laura and Anders Ferguson (2015). A New Pilar of Sustainability, Philantopic-Creativity, Foundation Center, New York. https://pndblo g.typepad.com/pndblog/2015/10/creativity-a-new-pillar-of-sustainabil ity.html.

Carson, Rachel, (1962). Silent Spring, A Mariner Book, Houghton Mifflin Company, Boston, New York.

Cheng, V. (2018). Views on Creativity, Environmental Sustainability and Their Integrated Development. Creative Education, 9, 719–743. doi: 10.4236/ce.2018.95054.

Clark, W. C. and Munn, R. E. (Eds.) (1986). Sustainable Development of the Biosphere, Cambridge: Cambridge University Press.

Colyvas, Jeannette A. and Walter W. Powell (2006). Roads to Institutionalization: The Remaking of Boundaries between Public and Private Science, Research in Organizational Behavior, 27, 305–353.

Crossan, M. and Bedrow, I. (2003). Organizational learning and strategic renewal. Strategic Management Journal, 24, 1087–1105.

Daly, H. E. (2007). Ecological Economics and Sustainable Development, Selected Essays of Herman Daly, Advances in Ecological Economics, MPG Books Ltd, Bodmin, Cornwall. http://library.uniteddiversity.coop/ Measuring_Progress_and_Eco_Footprinting/Ecological_Economics_and _Sustainable_Development-Selected_Essays_of_Herman_Daly.pdf.

DESA (2013). World Economic and Social Survey 2013, Sustainable Development Challenges, Department of Economic and Social Affairs, The Department of Economic and Social Affairs of the United Nations Secretariat, NY. https://sustainabledevelopment.un.org/content/do cuments/2843WESS2013.pdf.

Esty, K., Griffin, R. and Schorr-Hirsh, M. (1995). Workplace diversity. A manager's guide to solving problems and turning diversity into a competitive advantage. Avon, MA: Adams Media Corporation.

Fischer, D., Jenssen, S. and Tappeser, V. (2015). Getting an Empirical Hold of The sustainable University: A Comparative Analysis of Evaluation Frameworks across 12 Contemporary Sustainability Assessment Tools. Assess. Eval. High. Educ., 40, 785–800.

Fobes (2019). America's Wealth Inequality Is At Roaring Twenties Levels, Contributor-Jesse Colombo, Forbes (https://www.forbes.com/sites/jesse colombo/2019/02/28/americas-wealth-inequality-is-at-roaring-twenti es-levels/#62f244642a9c).

Freistein, K. and Mahlert, B. (2015). The Role of Inequality in the Sustainable Development Goals, Conference Paper, University of Duisburg-Essen See discussions, stats, and author profiles for this publication at: https://www.researchgate.net/publication/301675130.

Galbraith, James, K. (2012). Inequality and Instability: The Study of the World Economy Just before the Great Crisis. Oxford: Oxford University Press.

Genevey, R., Pachauri, R. K., Tubiana, L., Jozan, R., Voituriez, T. and Sundar, S. (2013). A planet for life: sustainable development in action, TERI, New Delhi http://regardssurlaterre.com/sites/default/files/edition/2016/01-PFL%202013%20Inequalities.pdf.

Girard, Luigi Fusco (2010). Sustainability, creativity, resilience: toward new development strategies of port areas through evaluation processes, Int. J. Sustainable Development, 13(1/2), 161.

Guidetti, Rehbein (2014). Theoretical Approaches to Inequality in Economics and Sociology, Transcience, 5(1), ISSN 2191-1150. https://www2.hu-berlin.de/transcience/Vol5_No1_2014_1_15.pdf.

Håvard Mokleiv Nygård (2017). Achieving the sustainable development agenda: The governance – conflict nexus, International Area Studies Review, 20(1), 3–18.

Holvino, E., Ferdman, B. M. and Merrill-Sands, D. (2004). Creating and sustaining diversity and inclusion in organizations: Strategies and approaches. In M. S. Stockdale and F. J. Crosby (Eds.). The psychology and management of workplace diversity. Malden, Blackwell Publishing, pp. 245–276.

IISD (2005). Indicators. Proposals for a way forward. Prepared L. Pinter, P. Hardi, P. Bartelmus. International Institute for Sustainable Development Sustainable Development, Canada Retrieved January 8, 2015, from https://www.iisd.org/pdf/2005/measure_indicators_sd_way_forward.pdf.

IISD (2013). The Future of Sustainable Development: Rethinking sustainable development after Rio+20 and implications for UNEP. International Institute for Sustainable Development Retrieved November 5, 2015, from http://www.iisd.org/pdf/2013/future_rethinking_sd.pdf.

IUCN (1980). World Conservation Strategy: Living Resource Conservation for Sustainable Development. Retrieved November 7, 2015, from https://portals.iucn.org/library/efiles/documents/WCS-004.pdf.

IUCN, UNDP and WWF (1991). Caring for the Earth. A Strategy for Sustainable Living. International Union for Conservation of Nature and Natural Resources, United Nations Environmental Program & World Wildlife Fund, Retrieved November 8, 2015, from https://portals.iucn.org/library/efiles/documents/CFE-003.pdf.

IUCN, UNDP and WWF (1991). Caring for the Earth. International Union for Conservation of Nature and Natural Resources, United Nations Environmental Program & World Wildlife Fund.

Jackson, Tim (2009). Prosperity without Growth: Economics for a Finite planet. Abingdon, United Kingdom: Earthscan.

Jan-Peter Voß, Jens Newig, Britta Kastens, Jochen Monstadt and Benjamin Nölting (2007). Steering for Sustainable Development: a Typology of Problems and Strategies with respect to Ambivalence, Uncertainty and Distributed Power. Journal of Environmental Policy & Planning, 9(3–4), 193–212. https://www.researchgate.net/profile/Jochen_Monstadt/publication/233049753_Steering_for_Sustainable_Development_A_Typology_of_Problems_and_Strategies_with_Respect_to_Ambivalence_Uncertainty_and_Distributed_Power/links/577ff29608ae5f367d370a97/Steering-for-Sustainable-Development-A-Typology-of-Problems-and-Strategies-with-Respect-to-Ambivalence-Uncertainty-and-Distributed-Power.pdf.

Jonathan M. Harris (2000). Basic Principles of Sustainable Development, Global Development and Environment Institute, Working Paper 00-04, Global Development and Environment Institute, Tufts University, https://tind-customer-agecon.s3.amazonaws.com/11dc38b4-a3e2-44d0

-b8c8-3265a796a4cf?response-content-disposition=inline%3B%20fil ename%2A%3DUTF-8%27%27wp000004.pdf&response-content-ty pe=application%2Fpdf&AWSAccessKeyId=AKIAXL7W7Q3XHXDV DQYS&Expires=1560578358&Signature=T%2BpMgFFZjQvmVL8 EHsy74Ds%2FKAM%3D.

Karlsson-Vinkhuyzen, Sylvia, Arthur L. Dahl and Asa Persson (2018). The emerging accountability regimes for the Sustainable Development Goals and policy integration: Friend or foe? Environment and Planning C: Politics and Space, 36(8), 1371–1390.

Klarin, Tomislav (2018). The Concept of Sustainable Development: From its Beginning to the Contemporary Issues, Zagreb International Review of Economics & Business, 21(1), 67–94. doi: https://doi.org/10.2478/zire b-2018-0005; https://content.sciendo.com/view/journals/zireb/21/1/art icle-p67.xml.

Lang, D. J., Wiek, A., Bergmann, M., Stauffacher, M., Martens, P., Moll, P., Swilling, M. and Thomas, C. J. (2012). Transdisciplinary research in sustainability science: practice, principles, and challenges. Sustainabil- ity Science, 7(1), 25–43.

Lélé Sharachchandra, M. (1991). Sustainable development: A critical review. World Development, 19(6), 607–621. https://edisciplinas.usp.br/plugin file.php/209043/mod_resource/content/1/Texto_1_lele.pdf.

Mair Simon, Aled Jones, Jonathan Ward, Ian Christie, Angela Druckman and Fergus Lyon (2017). A Critical Review of the Role of Indicators in Implementing the Sustainable Development Goals in the Handbook of Sustainability Science in Leal, Walter (Edit.) https://www.researchgate. net/publication/313444041_A_Critical_Review_of_the_Role_of_Indicat ors_in_Implementing_the_Sustainable_Development_Goals

Marin, C., Dorobantu, R., Codreanu, D. and Mihaela, R. (2012). The Fruit of Collaboration between Local Government and Private Partners in the Sustainable Development Community Case Study: County Valcea. Economy Transdisciplinarity Cognition, 2, 93–98.

Marshall, Alfred (1961), (originally 1920). Principles of Economics, 9th edn, New York: Macmillan.

Marx, Karl (1953ff). Marx-Engels-Werke (MEW). Berlin: Dietz.

Meadows, D. H. (1998). Indicators and Information Systems for Sustainable Development. A report to the Balaton Group 1998. The Sustainability Institute.

Meadows, D. H., Meadows, D. L., Randers, J. and Behrens III, W. W. (1972). The Limits of Growth. A report for the Club of Rome's project on the predicament of mankind. Retrieved September 20, 2015, from http://co llections.dartmouth.edu/published-derivatives/meadows/pdf/meadows_ ltg-001.pdf.

Nicolau, Melanie, Rudi W. Pretorius (2016). University of South Africa (UNISA): Geography at Africa's largest open distance learning institution, (in book). The Origin and Growth of Geography as a Discipline at South African Universities, Gustav Visser, Ronnie Donaldson, and Cecil Seethal, (eds), Stellenbosch, South Africa: Sun Press.

Norgaard, R. B. (1988). Sustainable development: A coevolutionary view, Futures, 20(6), 606–620.

North, Douglass C. (1981). Structure and Change in Economic History. New York: W.W. Norton & Co. Online: https://sustainabledevelopment.un. org/partnership/?p=11748 (accessed on 27 January 2019).

PAP/RAC (1999). Carrying capacity assessment for tourism development, Priority Actions Program, in framework of Regional Activity Centre Mediterranean Action Plan Coastal Area Management Program (CAMP) Fuka-Matrouh – Egypt, Split: Regional Activity Centre.

Pearce, D. (1989). Tourism Development. London: Harlow.

Purvis, Ben, Yong Mao and Darren Robinson (2018). Three pillars of sustainability: in search of conceptual origins, Sustainability Science, Springer, 23 https://doi.org/10.1007/s11625-018-0627-5r15%20-low%20res%20 20100615%20-.pdf.

Rockström, J., Steffen, W., Noone, K., Persson, Å, Chapin, S., Lambin, E., Lenton, T., Scheffer, M., Folke, C., Schellnhuber, H., Nykvist, B., de Wit, C., Hughes, T., van der Leeuw, S., Rodhe, H., Sörlin, S., Snyder, P., Costanza, R., Svedin, U., Falkenmark, M., Karlberg, L., Corell, R., Fabry, V., Hansen, J.,Walker, B., Liverman, D., Richardson, K., Crutzen, P. and Foley, J. (2009). A safe operating space for humanity. Nature, 461, 472–475.

Sayed, A. and Asmuss, M. (2013). Benchmarking Tools for Assessing and Tracking Sustainability in Higher Educational Institutions. Int. J. Sustain. High. Educ., 14, 449–465.

Scott, W. R. (2003). Institutional carriers: reviewing models of transporting ideas over time and space and considering their consequences. Industrial and Corporate Change, 12, 879–894.

Soares, Maria Clara Couto, Mario Scerri and Rasigan Maharajh (Eds.), (2014). Inequality and Development Challenges, Routledge, https://pr d-idrc.azureedge.net/sites/default/files/openebooks/032-9/.

Sterling, S. (2010). Learning for resilience, or the resilient learner? Towards a necessary reconciliation in a paradigm of sustainable education. Environmental Education Research, 16, 511–528, doi: 10.1080/13504622.2010.505427.

Tomislav Klarin (2018). The Concept of Sustainable Development: From its Beginning to the Contemporary Issues, Zagreb International Review of Economics & Business, 21(1), 67–94. doi: https://doi.org/10.2478/zire b-2018-0005 https://content.sciendo.com/view/journals/zireb/21/1/artic le-p67.xml.

UN, United Nations (1972). Report of the United Nations Conference on the Human Environment. Stockholm. Retrieved September 20, 2015 from http://www.un-documents.net/aconf48-14r1.pdf.

UN, United Nations (1997). Earth Summit: Resolution adopted by the General Assembly at its nineteenth special session. Retrieved November 4, 2015 from http://www.un.org/esa/earthsummit/index.html.

UN, United Nations (2002). Report of the World Summit on Sustainable Development, Johannesburg; Rio+10. Retrieved November 4, 2015 from http://www.unmillenniumproject.org/documents/131302_wssd_report_r eissued.pdf.

UN, United Nations (2010). The Millennium Development Goals Report. Retrieved September 20, 2015 from http://www.un.org/millenniumgoa ls/pdf/MDG%20Report%202010%20En%20.

UN, United Nations (2012). Resolution "the future we want". Retrieved November 5, 2015 from http://daccess-dds-ny.un.org/doc/UNDOC/ GEN/N11/476/10/PDF/N1147610.pdf?.

UN, United Nations (2015). Retrieved September 21, 2015 from http://www. un.org/en/index.html.

UN, United Nations (2015b). 70 years, 70 documents. Retrieved September 21, 2015 from http://research.un.org/en/UN70/about.

UN, United Nations (2015c). Resolution, Transforming our world: the 2030 Agenda for Sustainable Development. Retrieved November 5, 2015 from http://www.un.org/ga/search/view_doc.asp?symbol=A/RES/ 70/1&Lang=E.

UN, United Nations (2015d). The Millennium Development Goals Report 2015. Retrieved November 5, 2015, from http://www.un.org/millenniu mgoals/2015_MDG_Report/pdf/MDG%20.

UNDESA-DSD – United Nations Department of Economic and Social Affairs Division for Sustainable Development, 2002. Plan of Implementation of the World Summit on Sustainable Development:

The Johannesburg Conference. New York. UNESCO – United Nations Educational, Scientific and Cultural Organization, 2005. International Implementation Scheme. United Nations Decade of Education for Sustainable Development (2005–2014), Paris.

UNEP (2010). Background paper for XVII Meeting of the Forum of Ministers of Environment of Latin America and the Caribbean, Panamá City, Panamá, 26–30 April 2010, UNEP/LAC-IG.XVII/4, UNEP, Nairobi, Kenya, http://www.unep.org/greeneconomy/AboutGEI/WhatisGEI /tabid/29784/Default.aspx.

UNESCO (2013). UNESCO's Medium – The Contribution of Creativity to Sustainable Development Term Strategy for 2014–2021, http://www.un esco.org/new/fileadmin/MULTIMEDIA/HQ/CLT/images/CreativityFi nalENG.pdf

United Nations (2015). Transforming our world: the 2030 agenda for sustainable development. New York (NY): United Nations; 2015 (https://sus tainabledevelopment.un.org/post2015/transformingourworld, accessed 5 October 2015).

United Nations General Assembly (1987). Report of the world commission on environment and development: Our common future. Oslo, Norway: United Nations General Assembly, Development and International Co-operation: Environment.

Vander-Merwe, I. and Van-der-Merwe, J. (1999). Sustainable development at the local level: An introduction to local agenda 21. Pretoria: Department of environmental affairs and tourism.

Vos Robert, O. (2007). Perspective Defining sustainability: a conceptual orientation, Journal of Chemical Technology and Biotechnology, 82(1), 334–339.

Wals, A. (2009). United Nations Decade of Education for Sustainable Development (DESD, 2005–2014): Review of Contexts and Structures for Education for Sustainable Development Learning for a sustainable world. Paris.

WB, The World Bank (2015). World Development Indicators. Retrieved September 2, 2015, from http://data.worldbank.org/data-catalog/wo rld-development-indicators.

WCED (1987). Our Common Future World Commission on Environment and Development New York: Oxford University Press.

WID (2018). World Inequality Report 2018, The Paris School of Economics, Inequality Lab, WID. world, https://wir2018.wid.world/files/download/ wir2018-full-report-english.pdf.

Wiseman, Erica (2007). The Institutionalization of Organizational Learning: A Noninstitutional Perspective, Proceedings of OLKC 2007 – "Learning Fusion", UK, https://warwick.ac.uk/fac/soc/wbs/conf/olkc/archive/olkc 2/papers/wiseman.pdf.

Wright, E. O. (1994). Interrogating Inequalities. New York: Verso.

World Economic Outlook (2007). Globalization and Inequality, World Economic Forum, DC.

Wu, SOS, Jianguo (Jingle) (2012). Sustainability Indicators Sustainability Measures: Local-Level SDIs494/598. http://leml.asu.edu/Wu-SIs2015F/LECTURES+READINGS/Topic_08-Pyramid%20Method/Lecture-The%20Pyramid.pdf.

Yarime, M. and Tanaka, Y. (2012). The Issues and Methodologies in Sustainability Assessment Tools for Higher Education Institutions – A Review of Recent Trends and Future Challenges. J. Educ. Sustain. Dev., 6, 63–77.

2

Economic Growth and Regional Disparities: Literature Review in a Search for the Interconnections

Shvindina Hanna[1,2,*], Serhiy Lyeonov[3] and Tetyana Vasilyeva[4]

[1]Department of Management, Sumy State University, Ukraine
[2]Purdue University, USA
[3]Economic Cybernetics Department, Sumy State University, Ukraine
[4]Finance and Entrepreneurship Department, Sumy State University, Ukraine
E-mail: shvindina.hannah@gmail.com
*Corresponding Author

The study presented in this chapter is aimed to reveal the main problems in a sphere of economic growth and disparities globally. There are several Sustainable Development Goals that are linked to the problem of economic disparities and should be solved as soon as possible. The automated bibliometric tools for revealing the main trends and problems were used, and combination of these findings with inductive content-analysis made possible to state that there is no one solution for the existing problem. The problems coupled with economic growth are poverty reallocation, inequality, environmental degradation, climate change and risk. The review of academic literature was made to reveal the possible ways of further social and economic transformations.

2.1 Introduction

"The 2030 Agenda for Sustainable Development, adopted by all United Nations Member States in 2015, provides a shared blueprint for peace and prosperity for people and the planet, now and into

the future. At its heart are the 17 Sustainable Development Goals (SDGs), which are an urgent call for action by all countries – developed and developing – in a global partnership. They recognize that ending poverty and other deprivations must go hand-in-hand with strategies that improve health and education, reduce inequality, and spur economic growth – all while tackling climate change and working to preserve our oceans and forests"

(United Nations, 2015)

The major step in fostering sustainability begins from the first UN Conference on the Human Environment, Stockholm (1972), followed by the second Earth Summit in Rio de Janeiro, 1992, where 172, 108 people participated including head of the states, business personnel and other experts. In the first time about 2,400 representatives of non-governmental organizations (NGOs) participated in Rio summit. Summit produced agenda 21 declaration on environment and development, the statement of forest principles, the United Nations framework convention on climate change and the United Nations convention on biological diversity. Since Rio summit global concern on environment management and policy reform became common agenda to the entire world. Most of the states in the world started focus and monitor on patterns of production (i.e., toxic components, gasoline, and poisonous waste), investigation on alternatives for the fossil fuels (which is major cause for global climate change), alternatives for the public transportation (to reduce air pollution and smog) and water resource management (UNEP, 2010; UNESCO, 2013; United Nations, 2015; Bhandari, 2017, 2018). Blue print provides a comprehensive structure for the modernization of national/transnational environment protection and environment reform which includes the framework for sustainability and offers the links between economic growth using science and technology to solve the environmental problems with the application of multi-driven approaches. The world conferences based on sustainability and environmental reforms have been broadly focusing on the natural resource management, searching options to reduce the environment impact due to economic activities with the application of new technology (Bhandari, 2017, 2018).

"The Sustainable Development Goals (SDGs), otherwise known as the Global Goals, are a universal call to action to end poverty, protect the planet and ensure that all people enjoy peace and

prosperity" (UNDP, 2019). There are 17 SDGs and goal 16, states "Promote peaceful and inclusive societies for sustainable development, provide access to justice for all and build effective, accountable and inclusive institutions at all levels"

(UN, 2015; https://sustainabledevelopment.un.org/sdg16)

The Sustainable Development Goals (SDGs) is aimed to re-group and re-create the targets of global growth and development. The system of interrelated targets provides the indicators that enhancing the balanced co-governance through the collaboration between multiple stakeholders. At the same time despite the progress in SDGs implementation, there are some negative trends which should be investigated to be eliminated or minimized. One of the problems is increasing disparities in wages, labour market inequalities and energy consumption. Climate change remains the biggest challenge, and one of the consequences is its unequal distribution in the 'polluters – recipients' system. The countries that are leading in the energy consumption and environmental pollution (i.e., USA) are not the same countries that experience the biggest catastrophes, such as Philippines which was damaged by Super Typhoon Mangkhut in 2018. These trade-offs should be managed carefully, as well as anthropogenic disasters risks should be mitigated through pro-event and post-event strategies. This chapter is devoted to economic aspects of growth and disparities per se. The chapter is divided into two sub-chapters where the first part is devoted to bibliometric analysis of the new trends in a field of research; and the second half includes inductive content analysis of interrelated terms.

2.2 Literature Review and Bibliometric Analysis

A big number of reviews on sustainable development have been done in the past decades, starting from 90-ties (Lele, 1991; Mebratu, 1998) which were mostly focused on general framework of sustainable development. Later, the individual SDGs became an inspiration for the scientists' efforts, and each study had provided insights in a certain field, such as renewable energy (Dincer, 2000), green supply chain (Fahimnia et al., 2015; Eskandarpour et al., 2015). However, there are still many studies (for instance, Caiado et al., 2018) which review the SDG framework as a whole system. Our study is devoted to investigating the gaps and disproportions in the economic growth between the countries globally, and between outsiders and leaders to retrieve possible explanations of development disproportions.

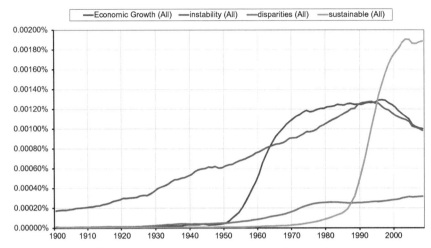

Figure 2.1   The frequency of keywords usage over the period of recent years.

Source: Google Ngram Tool.

Table 2.1   Publishing trends in a field of research

Search Keywords	No. of Papers
"economic growth" AND disparities AND instability AND sustainable	0
economic AND growth AND disparities AND instability AND sustainable	1
"economic growth" AND instability AND disparities	14
"economic growth" AND instability AND sustainable	96
"economic growth" AND sustainable AND disparities	**124**
"economic growth" AND disparities	1315
"economic growth" AND instability	1019
"economic growth" AND sustainable	8773

Source: Search results from SCOPUS Database.

The keywords used for data collection include: "Economic Growth", "Disparities", "Instability", "Sustainable". The combination of these keywords reveals interesting trends in publications (please see the Figure 2.1), where "Disparities" and "Sustainable" has become more popular recently, and "Economic Growth" and "Instability" have similar trends. We may state that insofar as scholar are interested in economic growth phenomenon, the instability of the growth got their attention inevitably.

Using the "title, abstract, keywords" search in Scopus database, we reveal slightly different trends in publishing, which are presented in a table below. Refining the keywords and their combinations in several steps of search we obtained the following data (please see Table 2.1).

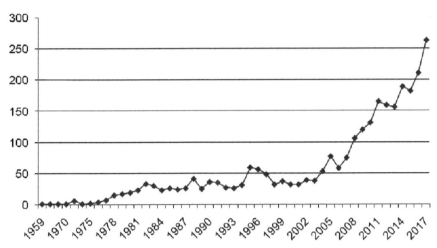

Figure 2.2 The publishing trend in using keyword "disparities".
Source: Constructed based on SCOPUS Database.

The only one paper which matches all the keywords is study done by Tian and his co-authors in 2017 (Tian et al., 2017). Refining keywords and search in different combinations let us to confirm big interest of scholars in the field of research in general and let us assume that studies in a sphere of regional disparities represent a new emerging trend in science. And to prove that thought, the number of papers with the keyword "disparities" is represented below where publishing trend is increasing (see Figure 2.2). The further bibliometric analysis can be performed by restoring the search results in RIS format that include all essential information (paper title, names of authors, affiliations, abstract, keywords and so on). The data set for further investigation is chosen for 124 papers in a field of research (highlighted in bold in Table 2.1).

Further data analysis includes Bibliometric Analysis using BibExcel. BibExcel is chosen due to flexibility of using different types of data (Persson et al., 2009), comparative simplicity of the data processing, and free access to the program. Thus, using BiblExcel, GPSVisualizer and GoogleMaps the contributors by the countries can be presented in a map (please see Figure 2.3) for chosen sample.

As we may see from that analysis recent papers were published by the scholars all around the world, but mostly from China, India, USA and Europe. It's partly can be explained by the selectivity of SCOPUS Database, but at

Figure 2.3   Geographical locations of contributing organizations.

Source: Authors (Constructed for N = 124, using SCOPUS database).

the same time it may provide interesting insights for further argumentations of economic growth and disparities studies.

2.3 The Content Analysis of Previous Findings: Interconnections of the Terms

VOSviewer is proven to be effective tool for providing visualization (van Eck and Waltman, 2013) of the keywords and terms usage, as well as for understanding of interconnections between main terms in the field of research. Using the abstracts of the chosen papers, the network of terms was generated. The terms are divided into clusters which are organized by association principle, identified by CitNetExplorer's clustering technique offered by VOSviewer program (van Eck and Waltman, 2013). As it is shown (please see Figure 2.4) there are three clusters (red, blue and green) that embrace 51 terms used most frequently in the chosen studies. Details items and clusters content are presented in Appendix (please see Table A1).

The cluster analysis brings not only visualization of terms using (i.e., frequency and density) but the information about interlinks between terms. For further understanding of economic development complexity the inter-linked items were identified and organized into two big groups, distinguished by authors – negative and semi-neutral. Accordingly, the semi-neutral items related to "economic development" are "competitiveness", "employment", "energy consumption", "GDP", "productivity", "regional development", "regional level", "rural area", "society", "state", "urbanization". As we see,

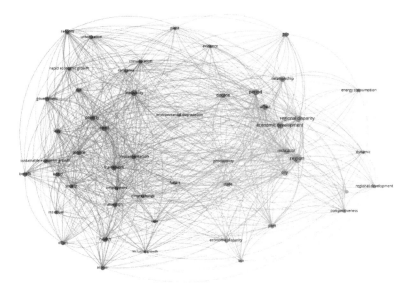

Figure 2.4 Network visualization of most frequent terms.
Source: Authors (Constructed via VOSViewer v.1.6.10 for the sample N = 124, f (*frequency*) ≥ 7).

semi-neutral means that terms can have both natures, as positive, as negative according to the meaning that authors put into the mentioned terms. For instance, competitiveness can be high or low, urbanization can have positive and negative outcomes. In the case of other group of items, they are clearly negative as far as economic development is coupled to "climate change", "economic disparity", "environmental degradation", "inequality", "lack [*of resources – authors*]", "poverty", "risk". Other items, such as "indicator", "evidence", "case", "framework" and so on were excluded out of further investigation as insignificant or too general.

If we compare the "economic development" network and "economic disparities" network (please see Figure B1 in appendix), we may see that mentioned terms are overlapped.

These findings became the inputs for further considerations of contradictories of economic growth and interconnections between main terms.

2.4 Economic Growth and Equality

Over the past decades trade-offs between economic growth and equality have been in focus of researchers. According to the results of search via Harzing's Publish or Perish v.6, in a period 1990–2010 the most cited publications belongs to Alesina and Rodrik (1994), Wilkinson (2002), Barro (1996), Galor and Zeira (1993), Pickett and Wilkinson (2010). Since 2011, the most contributing authors in terms of citations are Piketty (2015), Stiglitz (2012), Balassa (2013), Barr (2012), Bartels (2016), Wilkinson and Pickett (2011), Sherraden and Gilbert (2016), Atkinson (2015), Flora (2015), Corak (2013), Ostry and his co-authors (Ostry et al., 2014), Milanovic (2009, 2016). If Milanovic (2009) made an analysis of inequalities between countries (or global inequality), recent study of Cingano (2014) is focused on the inequalities between social classes in OCED countries, for instance, according to his research, the gap between rich and poor is highest for the recent 30 years. Moreover, Cingano found out that inequality, in its turn, affects economic growth by undermining education opportunities for children from poor socio-economic backgrounds, lowering social mobility and hampering skills development (Cingano, 2014).

These and mentioned above studies emphasized on income or/and wealth inequality, while the gender inequality remains a significant issue. The gender experts in national and international policies implementation found the achievements in the implementation of gender mainstreaming quite disappointing (Parpart, 2014). The term 'gender mainstreaming' has become well-spread, promising equality and women's empowerment into institutions, programs and policies. Being reachable at the first sight, these goals wrapped into the indicators associated with the MDG 3, are insufficient (Naila Kabeer, 2005). The expert argues that transformations can be achieved through the education, which enable women to take part in decision-making and increase access to resources within and beyond the household (Naila Kabeer, 2005).

Indeed, the problems of wealth, income and gender inequality are interrelated, and linked to access to higher education. But even with improved situation about access to education and college experience (Jacobs, 1996), the post-educational outcomes differ, and women are still behind men in wages and salaries. According to the research of Blau and Kahn (2017), the trends in graduation from HEIs have changed since 70-ties, when men were more likely than women to go to college and beyond. In 2011, USA women earned 57 per cent of bachelor's degrees and 62 per cent of associate degrees. There have been comparable gains at the post-graduate level – with women receiving 61 per cent of master's degrees, 51 per cent of Ph.D.'s and 49 per cent of first professional degrees in United States. The same trend was observed in advanced nations and many developing countries across the world (Blau and Kahn, 2017). In general, in USA the gender wage gap was 24% in 1980, and 16% in 2010 (Blau and Kahn, 2017), while in Europe the gender gap was 12.5% in 2014 (Redmond and Mcguinness, 2019). According to Global Wage Report (2018), the gender wage gap remains unacceptably high, about 20 per cent globally, which represents "one of today's the greatest manifestations of the social injustice", as mentioned Guy Ryder in introduction to the report (Global Wage Report, 2018), and the differences in the levels of education between men and women play a limited role in explaining gender gaps.

Another big issue related to income inequality and economic disparities is inequities in health care services. In the study of Devaux (2015) the data from 18 countries in the period of 2006–2009 were analyzed and proved that income-related inequalities and inequities in health care service utilization still exist in OECD countries. In most countries, for the same needs for health care, people with higher incomes are more likely to consult a doctor, than those who are with lower incomes (Devaux, 2015). But at the same time, part of these cross-country disparities can be explained by the differences in health system, insurance policy, health care financing system, and so on. But income or wealth-related access to the health service remains big problem in modern society and should be solved within sustainable development framework.

Later research for 14 OPEC member countries (Vasylieva et al., 2018a; Abaas et al., 2018), proved that economic growth rates are related to a quality of social sector, in particularly an improvement in life expectancy by one per cent leads on average to 0.5–1.33% growth in GDP per capita (Abaas et al., 2018).

Anti-poverty programs are not enough to solve the inequality problems in a sphere of gender wage gap, age-related pay gap and inequalities in health care services. High-quality formal and informal education, training, coaching, advocacy projects and advising, in our opinion, may become powerful drivers of transformations. High or growing inequalities and disparities between countries, regions, social classes are critical in a way to sustained growth, and need to be more investigated, and at the same time to be immediately mitigated. Meanwhile, the investments into education stay underestimated by low-income population, which leads to a poverty loop. In our opinion, the cooperation between NGOs, government agencies and HEIs will allow to overcome these stereotypes.

2.5 Economic Growth and Stability (or Risk)

In his seminal work Stiglitz (2000) while analyzing Asian financial crisis of 1997, mentioned that crises spread very fast and affected countries with good economic policies and seriously. Clearly, later financial crisis of 2008 confirmed the thought of Stiglitz that crises become more frequent and more severe. Indeed, the war conflict in Ukraine started in 2012 by Russian Federation, the Brexit that had been widely discussed since 2016 can be considered as the long-lasting crises in Europe. If Ukraine is not comparable to UK in terms of country wealth or GDP per capita, in terms of populations size and number of habitants affected by the crises these countries are fellow sufferers. Political instability can be barrier for economic growth which was investigated widely and the negative relation between political instability and investments was proved (i.e., Alesina et al., 1996), as well as between political instability and productivity growth (Aisen and Veiga, 2013). Other findings should be considered about positive correlation between per capita income and sociopolitical destabilization in modernizing low-and middle-income economies (Korotayev et al., 2018). For instance, per capita income growth in authoritarian regimes tends to lead to an intensification of democratic movements and certain destabilization of the system (Korotayev et al., 2018). Indeed, political instability is only one of risk factors for economic growth. We should also mention studies that proved significant impact of fiscal decentralization measures on different aspects of socio-economic development of the countries (Chygryn et al., 2018), and the dependency of country income on macroeconomic stability (Lyeonov et al., 2018; Djalilov et al., 2015). The studies on proving the links between economic growth of the country and macroeconomic and political stability are in a center of academic attention (i.e., Vasylieva et al., 2018b). Especially, considering that group of scientists proved that effective management of innovations lead to sustainable economic growth (Kendiukhov et al., 2017). Leaving this issue for other researchers to investigate, we'd like to shed some light on existing studies in economic and environmental instability.

In our opinion, it is important to understand that economic growth does not guarantee stability or security, moreover rapid growth under certain conditions means turbulence for industry players and markets. There are growth-instability combinations presented by Chandra (2003) in his research. Despite the fact that data used in this study were computed for the period 1980–1995 (sixteen years), before two big financial crises, this empirical analysis of the frontier provided a realistic framework for understanding of importance of sectorial structure for economic outcomes, and optimal structure that can be modified through industrial policy towards growth and stability. Stability and instability are highly interrelated in the academic literature. In one of our earlier papers (Lyulyov and Shvindina, 2017) the analysis of existing instability models was presented, where Macroeconomic Stabilization Pentagon Model offered by Kolodko was chosen to re-construct the profiles of main European countries. As a result, some paradoxes were revealed, for instance, one of them we can formulate as *"stability does not mean economic growth neither"*, meaning that Belarus (post-Soviet country) under crises of 2007–2008 was the most stabilized among EU and non-EU countries. At the same time the standards of living in Belarus are low comparatively to EU and other post-Soviet countries (Latvia, Lithuania and Estonia, for example). Therefore, we may assume, that there are different combinations of the stability-growth parameters that should be investigated further.

2.6 Economic Growth and Climate Change

All sorts of research, reports, studies and data were published on the topic of climate change in recent decades. For the rest 5 years among all the research we should mention top-cited results on air pollution (Seinfeld and Pandis, 2016), detailed report on climate change that united researchers all around the world (IIPC, 2013), modelling of different scenarios applied for global greenhouse gas emission (Rogelj et al., 2016), on understanding of the permafrost carbon feedback on climate (Schuur et al., 2015), regional climate modelling for different communities of Europe (Jacob et al., 2013).

Climate change is closely linked to urbanization and population growth, as well as economic growth which has urbanization as a core element. The similar conclusion was made by Dkhili (2018). By 2050, by experts' estimations, the world's urban population is expected to nearly double, and over 90 per cent of this growth will take place in Africa, Asia, Latin America and the Caribbean (Habitat III, 2016). And from the evidences we may approve that urban economy makes up two-thirds of the current global economy and in soon future will cross three-fourths. Geographically, at the same time urbanization will take place in developed countries rather than go across Asia and Africa, but ecological consequences of climate changes will be shared globally, which is disproportional to the 'contributions. Ecologically, urban consumption is the danger for net primary production of the biosphere and putting most countries in ecological deficit since the 1980s, according to (Revi, 2016).

One more interesting phenomenon exist which we may name as 'environmental disparities' which means that the inputs and outputs into environment degradation and climate change are distributed unfair. The polluters of CO_2 which are mostly developed countries are not the main recipients of the consequences. For instance, Philippines which was damaged by Super

Typhoon Mangkhut in 2018, was actually damaged by climate change effect. Meanwhile the biggest contributors into CO_2 emissions are China, the USA, India, Russia and Japan. These trade-offs should be managed carefully, as well as anthropogenic disasters risks should be mitigated through pro-event and post-event strategies. At the same time, the study of Bilan and his co-authors (2019) highlighted that transition towards of renewable energy leads to decreasing of the CO_2 emissions and simultaneously increasing of economic growth (measured by GDP).

Summarizing the above mentioned in this paragraph, we can say that economic growth is a paradoxical phenomenon, which, on the one hand, brings its fruits to the improvement of people's living standards (urbanization, for example), and on the other hand, it causes environmental degradation. Sustainable development is a somewhat idealistic concept that enables to balance the momentary gains of economic growth and future losses from environmental consequences.

Establishing monitoring mechanisms of that kind is one of the main priorities for SDGs, according to (Lu et al., 2015), and we agree with that, especially considering that report on the SDG progress (Brackley and York, 2019) include survey data, not indicators evaluation. To monitor changes for further balancing is another challenge in a sphere of implementation of SDGs policy. It's crucial to formulate a comprehensive and balanced methodology of tracking values or indicators for further policymaking. And thereby there are a lot of suggestions which indicator should be tracked – energy or water consumption, emissions, health impacts, CO_2 or metal concentration, or water pH?

2.7 Economic Growth and Poverty

The PovcalNet data, presented below, shows the distribution of the world population across different poverty thresholds and represents the falling of extreme poverty (please see Figure 2.3). But at the same time there is still a big challenge in terms of economic disparities solution, as two-thirds of global population live using less than 10 dollars per day. The contribution into the poverty falling was done due to economic growth in recent decades. The combination of the poverty data and GDP per capita (see Figures 2.5 and 2.6) allows to see the trends of prosperity reallocation.

It is obvious that progress does not take place with the same velocity everywhere, and therefore economic growth creates inequalities between regions. Historical data proves that every region is richer than earlier in its

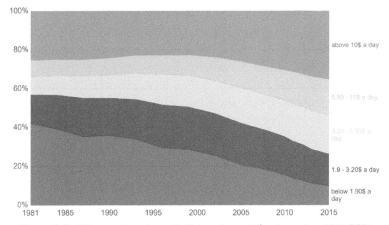

Figure 2.5 The number of people living above 10$ a day using 2011 PPP.

Source: World Bank, PovcalNet, presented in (Roser, 2019).

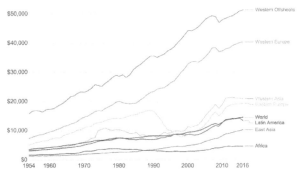

Figure 2.6 Average GDP per capita across countries and regions (adjusted for price changes over time, measured in international $ in 2011 prices).

Source: World Bank, presented in Jutta et al., 2018.

own history, though economy of the regions differ in terms of productivity and resources inputs. The economic disparities were extremely high in pre-modern societies (Jutta et al., 2018; Roser, 2019), but now when global GDP per capita becomes an indicator for world economy, there is a suggestion that world economy can become positive-sum game, where players may win without taking an advantage of other players. If earlier the war took place mostly because of resources conflicts, nowadays for economy growth a resources input may matter less than productivity. The Singapore phenomenon, Japanese economy, rapid GDP growth of India are just several cases that can become an illustration for policy implementation in a sphere of the economic disparities solution. Global Competitiveness Index (GCI) is a framework for measuring success of economies which is a system that integrates macroeconomic and microeconomic factors of competitiveness (Porter et al., 2008). In total there are 98 indicators organized into 12 pillars to measure competitiveness (Schwab, 2018). It works well to evaluate the progress in competitiveness and economic growth of different countries, individually and in comparison, between regions. In our opinion, GCI can be adjusted to SDG framework for further measurement and evaluation of sustainable development progress.

2.8 Conclusions and Discussion

This study combines automated bibliometric analysis methods and inductive literature review and brings together the terms coupled to economic growth and economic disparities. Among all the key words revealed in a process of cluster analysis, authors focused their attention on "climate change", "environmental degradation", "inequality", "poverty" and "risk" (instability in our edition). There are many phenomena remained unstudied, for instance, the interlinks between economic growth and health care indicators (mortality, morbidity and income-related issues), the influence of urbanization on economic growth which is investigated in numbers by Roser (2019), or employment, and so on. At the same time, there is a discussion in academic literature on Malthusian trap. Representatives of one flow suggested that economies of modern era broke out this trap due to technological innovations (Roser, 2019), other researchers insist that we are still in it – decline of death rates, urbanization, and growth of proportions of young cohorts in the overall population creates new trap after escaping from Malthusian one (Korotayev et al., 2018). Another debate exists in a sphere of technological readiness and productivity paradox. The productivity is a driver for economic growth but increasing investments into technologies are not leading to productivity increasing. One of the explanations is a shrinking life cycle of product implementation in the global markets. The choice of indicator for economic growth evaluation is needed to be investigated more, for instance GCI can be adjusted to Economic growth and SDGs indicators and that will leave the room for comprehensive benchmarking between regions and countries.

There are still the question of scale and scope of economic and social cohesion – should it take place between regions, between countries or globally? The movement in this sphere will certainly impact the economic disparities problem.

Acknowledgements and Funding

This research was supported through a 2018–2019 Fulbright Scholar Award. This research was funded by the grant from the Ministry of Education and Science of Ukraine (№ g/r 0118U003569).

We thank our colleague from Purdue University, Professor Vincent Duffy (School of IE, Purdue University) for his comments and inputs that greatly improved the paper.

Appendix A

Table A1 Terms Clusters for revealed items (Constructed via VOSViewer v.1.6.10, No. of papers = 124, No. of terms = 51)

Cluster 1 (Red)	Cluster 2 (Green)	Cluster 3 (Blue)
access	city	decade
attention	competitiveness	economic development
benefit	dynamic	economic disparity
case	indicator	effect
climate change	lack	energy consumption
consideration	part	environmental degradation
effort	productivity	evidence
employment	region	future
framework	regional development	GDP
government	regional disparity	period
health	regional level	place
implementation	state	relationship
inclusive growth	**Total (12 items)**	**Total (12 items)**
individual		
inequality		
need		
poverty		
rapid economic growth		
reform		
risk		
rural area		
service		
society		
sustainable economic growth		
urbanization		
way		
world		
Total (27 items)		

Appendix B

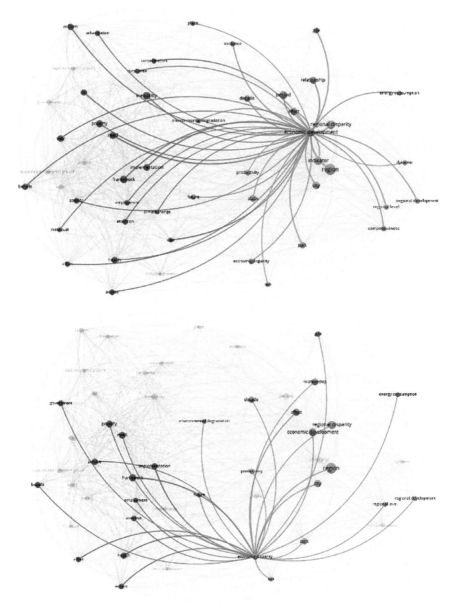

Figure B1   Comparison of "economic development" and "economic disparities" networks. (Constructed via VOSViewer v.1.6.10 for the sample of papers N = 124, number of items = 51).

References

Abaas, M. S. M., Chygryn, O., Kubatko, O. and Pimonenko, T. (2018). Social and economic drivers of national economic development: The case of OPEC countries. Problems and Perspectives in Management, 16(4), 155–168. doi: 10.21511/ppm.16(4).2018.14.

Aisen, A. and Veiga, F. J. (2013). How does political instability affect economic growth? European Journal of Political Economy, 29, 151–167.

Aiyar, S. S. and Ebeke, C. (2019). Inequality of Opportunity, Inequality of Income and Economic Growth.

Alesina, A. and Rodrik, D. (1994). Distributive politics and economic growth. The Quarterly Journal of Economics, 109(2), 465–490.

Alesina, A., Özler, S., Roubini, N. and Swagel, P. (1996). Political instability and economic growth. Journal of Economic growth, 1(2), 189–211.

Allen, C., Metternicht, G. and Wiedmann, T. (2016). National pathways to the Sustainable Development Goals (SDGs): A comparative review of scenario modelling tools. Environmental Science & Policy, 66, 199–207.

Atkinson, A. B. (2015). Inequality. Harvard University Press.

Balassa, B. (2013). The theory of economic integration (Routledge revivals). Routledge.

Barr, N. (2012). Economics of the welfare state. Oxford university press.

Barro, R. J. (1996). Determinants of economic growth: a cross-country empirical study (Working Paper No. 5698). National Bureau of Economic Research.

Bartels, L. M. (2016). Unequal democracy: The political economy of the new gilded age. Princeton University Press.

Bhandari, M. P. (2017). Climate change science: a historical outline. Adv Agr Environ Sci., 1(1), 1–8: 00002. http://ologyjournals.com/aaeoa/aaeoa_00002.pdf.

Bhandari, Medani, P. (2018). Green Web-II: Standards and Perspectives from the IUCN, Published, sold and distributed by: River Publishers, Denmark/the Netherlands. ISBN: 978-87-70220-12-5 (Hardback) 978-87-70220-11-8 (eBook).

Bilan, Y., Streimikiene, D., Vasylieva, T., Lyulyov, O., Pimonenko, T. and Pavlyk, A. (2019). Linking between renewable energy, CO_2 emissions and economic growth: Challenges for candidates and potential candidates for the EU membership. Sustainability (Switzerland), 11(6). doi: 10.3390/su11061528.

Blau, F. D. and Kahn, L. M. (2017). The gender wage gap: Extent, trends and explanations. Journal of Economic Literature, 55(3), 789–865.

Bracklye, A. and York, B. (2019). Evaluating Progress on the SDGs. GlobeScan-SustainAbility Survey. Retrieved from: http://sustain ability.com/wp-content/uploads/2019/03/globescan-sustainability-sdgs-survey-2019.pdf [Online Resource].

Caiado, R. G. G., Leal Filho, W., Quelhas, O. L. G., de Mattos Nascimento, D. L. and Ávila, L. V. (2018). A literature-based review on potentials and constraints in the implementation of the sustainable development goals. Journal of cleaner production, 198, 1276–1288.

Chandra, S. (2003). Regional economy size and the growth–instability frontier: Evidence from Europe. Journal of Regional Science, 43(1), 95–122.

Chygryn, O., Petrushenko, Y., Vysochyna, A. and Vorontsova, A. (2018). Assessment of fiscal decentralization influence on social and economic development. Montenegrin Journal of Economics, 14(4), 69–84. doi: 10.14254/1800-5845/2018.14-4.5.

Cingano, F. (2014). Trends in Income Inequality and its Impact on Economic Growth, OECD Social, Employment and Migration Working Papers, No. 163, OECD Publishing, Paris. https://doi.org/10.1787/5jxrjncw xv6j-en [Online Resource].

Corak, M. (2013). Income inequality, equality of opportunity and intergenerational mobility. Journal of Economic Perspectives, 27(3), 79–102.

Devaux, M. (2015). Income-related inequalities and inequities in health care services utilisation in 18 selected OECD countries. The European Journal of Health Economics, 16(1), 21–33.

Dincer, I. (2000). Renewable energy and sustainable development: a crucial review. Renewable and sustainable energy reviews, 4(2), 157–175.

Dkhili, H. (2018). Environmental performance and institutions quality: evidence from developed and developing countries. Marketing and Management of Innovations, 3, 333–244. http://doi.org/10.21272/mmi.2018.3-30.

Djalilov, K., Lyeonov, S. and Buriak, A. (2015). Comparative studies of risk, concentration and efficiency in transition economies. Risk governance and control: financial markets & institutions, 5(4 Continued-1), 178–187.

Eskandarpour, M., Dejax, P., Miemczyk, J. and Péton, O. (2015). Sustainable supply chain network design: an optimization-oriented review. Omega, 54, 11–32.

Fahimnia, B., Sarkis, J. and Davarzani, H. (2015). Green supply chain management: A review and bibliometric analysis. International Journal of Production Economics, 162, 101–114.

Flora, P. (2017). Development of welfare states in Europe and America. Routledge. doi: https://doi.org//10.4324/9781351304924.

Galor, O. and Zeira, J. (1993). Income distribution and macroeconomics. The review of economic studies, 60(1), 35–52.

Global Wage Report 2018/19 (2018): What lies behind gender pay gaps. Retrieved from: http://www.bpw.fr/files/3415/4326/9389/wcms_650 553.pdf [Online Resource].

Habitat III (2016). New urban agenda. Quito declaration on sustainable cities and human settlements for all. Quito UN Habitat. Retrieved from: http://habitat3.org/wp-content/uploads/Habitat-III-New-Urban-Agenda-10-September-2016.pdf [Online Resource].

IIPC (2013). Climate Change 2013: The Physical Science Basis. Contribution of Working Group I to the Fifth Assessment Report of the Intergovernmental Panel on Climate Change (Stocker, T. F., Qin, D., Plattner, G. K., Tignor, M., Allen, S. K., Boschung, J., Nauels A., Xia Y., Bex V. and Midgley, P. M. (eds.)) Cambridge University Press, Cambridge, United Kingdom and New York, NY, USA, 1535 pp.

Jacob, D., Petersen, J., Eggert, B., Alias, A., Christensen, O. B., Bouwer, L. M. and Georgopoulou, E. (2014). EURO-CORDEX: new high-resolution climate change projections for European impact research. Regional environmental change, 14(2), 563–578.

Jacobs, J. A. (1996). Gender inequality and higher education. Annual review of sociology, 22(1), 153–185.

Jutta, B., Inklaar, R., de Jong, H. and van Zanden, J. L. (2018). Project Database, version 2018. Rebasing 'Maddison': new income comparisons and the shape of long-run economic development, Maddison Project Working paper 10. Retrieved from: https://www.rug.nl/ggdc/hist oricaldevelopment/maddison/releases/maddison-project-database-2018 and http://www.ggdc.net/maddison/oriindex.htm [Online Resource].

Kabeer, N. (2005). Gender equality and women's empowerment: A critical analysis of the third millennium development goal 1. Gender and Development, 13(1), 13–24.

Kendiukhov, I. and Tvaronaviciene, M. (2017). Managing innovations in sustainable economic growth. Marketing and Management of Innovations, 3, 33–42.

Korotayev, A., Vaskin, I., Bilyuga, S., Khokhlova, A., Baltach, A., Ivanov, E. and Meshcherina, K. (2018). Economic Development and Sociopolitical Destabilization: A Re-Analysis. Cliodynamics, 9, 59–118.

Lele, S. M. (1991). Sustainable development: a critical review. World development, 19(6), 607–621.

Lyulyov, O. and Shvindina, H. (2017). Stabilization Pentagon Model: Application in the Management at Macro- and Micro-levels. Problems and Perspectives in Management, 15(3), 42–52.

Lyeonov, S. V., Vasylieva, T. A. and Lyulyov, O. V. (2018). Macroeconomic stability evaluation in countries of lower-middle income economies. Naukovyi Visnyk Natsionalnoho Hirnychoho Universytetu, 138–146. doi: 10.29202/nvngu/2018-1/4.

Lu, Y., Nakicenovic, N., Visbeck, M. and Stevance, A. S. (2015). Policy: Five priorities for the UN sustainable development goals. Nature News, 520(7548), 432.

Mebratu, D. (1998). Sustainability and sustainable development: historical and conceptual review. Environmental impact assessment review, 18(6), 493–520.

Milanovic, B. (2009). Global inequality and the global inequality extraction ratio: the story of the past two centuries. The World Bank.

Milanovic, B. (2016). Global inequality: A new approach for the age of globalization. Harvard University Press.

Moss, N. E. (2002). Gender equity and socioeconomic inequality: a framework for the patterning of women's health. Social science and medicine, 54(5), 649–661.

Ostry, M. J. D., Berg, M. A. and Tsangarides, M. C. G. (2014). Redistribution, inequality and growth. International Monetary Fund.

Parpart, J. L. (2014). Exploring the transformative potential of gender mainstreaming in international development institutions. Journal of International Development, 26(3), 382–395.

Persson, O., Danell, R. and Schneider, J. W. (2009). How to use Bibexcel for various types of bibliometric analysis. Celebrating scholarly communication studies: A Festschrift for Olle Persson at his 60th Birthday, 5, 9–24.

Pickett, K. and Wilkinson, R. (2010). The spirit level: Why equality is better for everyone. Penguin UK.

Piketty, T. (2015). About capital in the twenty-first century. American Economic Review, 105(5), 48–53.

Porter, M. E., Delgado, M., Ketels, C. and Stern, S. (2008). Moving to a new global competitiveness index. The global competitiveness report, 43–63.

Redmond, P. and Mcguinness, S. (2019). The Gender Wage Gap in Europe: Job Preferences, Gender Convergence and Distributional Effects. Oxford Bulletin of Economics and Statistics, 81(3), 564–587.

Revi, A. (2016). Afterwards: Habitat III and the Sustainable Development Goals. Urbanisation, 1(2), x–xiv.

Rogelj, J., Den Elzen, M., Höhne, N., Fransen, T., Fekete, H., Winkler, H. and Meinshausen, M. (2016). Paris Agreement climate proposals need a boost to keep warming well below 2 C. Nature, 534(7609), 631.

Roser, M. (2019). Economic Growth. Published online at OurWorldInData.org. Retrieved from: https://ourworldindata.org/economic-growth [Online Resource].

Schuur, E. A., McGuire, A. D., Schädel, C., Grosse, G., Harden, J. W., Hayes, D. J. and Natali, S. M. (2015). Climate change and the permafrost carbon feedback. Nature, 520(7546), 171.

Seinfeld, J. H. and Pandis, S. N. (2016). Atmospheric chemistry and physics: from air pollution to climate change. John Wiley and Sons.

Sherraden, M. and Gilbert, N. (2016). Assets and the poor: New American welfare policy. Routledge.

Schwab, K. (2018). The global competitiveness report 2017–2018. Geneva: World Economic Forum. Retrieved from: http://www3.weforum.org/doc s/GCR2018/05FullReport/TheGlobalCompetitivenessReport2018.pdf [Online Resource].

Solt, F. (2009). Standardizing the world income inequality database. Social Science Quarterly, 90(2), 231–242.

Stiglitz, J. E. (2000). Capital market liberalization, economic growth and instability. World development, 28(6), 1075–1086.

Stiglitz, J. E. (2012). The price of inequality: How today's divided society endangers our future. WW Norton and Company.

Tian, X., Geng, Y., Viglia, S., Bleischwitz, R., Buonocore, E. and Ulgiati, S. (2017). Regional disparities in the Chinese economy. An emergy evaluation of provincial international trade. Resources, Conservation and Recycling, 126, 1–11.

UNEP (2010). Background paper for XVII Meeting of the Forum of Ministers of Environment of Latin America and the Caribbean, Panamá City, Panamá, 26–30 April 2010, UNEP/LAC-IG.XVII/4, UNEP, Nairobi, Kenya.

UNESCO (2013). UNESCO's Medium-The Contribution of Creativity to Sustainable Development Term Strategy for 2014–2021, http://www.unesco.org/new/fileadmin/MULTIMEDIA/HQ/CLT/images/Creativity FinalENG.pdf.

United Nations (2015). Transforming our world: the 2030 agenda for sustainable development. New York (NY): United Nations; 2015 (https://sustainabledevelopment.un.org/post2015/transformingourworld, accessed 5 October 2015).

United Nations Educational, Scientific and Cultural Organization (UNESCO) (2009). Bonn Declaration. UNESCO World Conference on Education for Sustainable Development, Bonn, Germany, 31 March to 2 April 2009. http://www.esd-world-conference-2009.org/fileadmin/download/ESD2009BonnDeclaration080409.pdf.

United Nations General Assembly (1987). Report of the world commission on environment and development: Our common future. Oslo, Norway: United Nations General Assembly, Development and International Co-operation: Environment.

van Eck, N. J. and Waltman, L. (2013). VOSviewer manual. Leiden: Univeristeit Leiden, 1(1).

Vasilyeva, T., Lyeonov, S., Adamièková, I. and Bagmet, K. (2018a). Institutional quality of social sector: The essence and measurements. Economics and Sociology, 11(2), 248–262. doi: 10.14254/2071-789X.2018/11-2/17.

Vasylieva, T., Lyeonov, S., Lyulyov, O. and Kyrychenko, K. (2018b). Macroeconomic stability and its impact on the economic growth of the country. Montenegrin Journal of Economics, 14(1), 159–170. doi:10.14254/1800-5845/2018.14-1.12.

Wilkinson, R. G. (2002). Unhealthy societies: the afflictions of inequality. Routledge.

Wilkinson, R. and Pickett, K. (2011). The spirit level: Why greater equality makes societies stronger. Bloomsbury Publishing, USA.

3

Overcoming Inequalities as A Source of Economic Development

Zharova Liubov

Wyższa Szkoła Ekonomiczno-Humanistyczna/The University of Economics and Humanities, Bielsko-Biala, Poland
E-mail: zharova.l@gmail.com

Analysis of contemporary approaches to economic development and backgrounds of the success of developed countries. The evolution of factors of growth through changing the understanding of developing's priorities (switching from economic to social and environmental aspects). Why equality matters. The concept of "re-thinking" of modern management approaches. Generalizing of ways of overcoming common barriers of economic development.

3.1 Introduction

3.1.1 The Movement From Economic Growth to Economic Development

Sustainable development is the commonly used phrase that sometimes just the sign of "being in trend". Nevertheless, we believe that it is more than that even and we cannot avoid the ambition of finding the balance. We have stated that sustainable development it is a framework for contemporary economic development.

To disclose the statement, we should clarify the definitions because it is ensured that we will discuss the same things. The most known definition of sustainable development, that can be found in Our Common Future (1987). Sustainable development is development that meets the needs of the present without compromising the ability of future generations to meet their own needs. It contains within it two key concepts:

- The concept of 'needs', in particular the essential needs of the world's poor, to which overriding priority should be given; and
- The idea of limitations imposed by the state of technology and social organization on the environment's ability to meet present and future needs.

Interesting fact that symptoms and causes of threatened situation in the world in 1987 named poverty, growth, survival and the economic crisis. Actually, they still the same, analysis of the current situation also includes issues of poverty, survival, growth, and crisis (the new possible wave of crisis is forecast and discussing again).

Sustainability is the foundation for today's leading global framework for international cooperation – the Agenda 2030 (2015) for sustainable development and its Sustainable Development Goals (SDGs). The Goals and targets will stimulate action over the next 15 years in areas of critical importance for humanity and the planet. These 5Ps for sustainable development are:

People – this is about ending poverty and hunger, in all their forms and dimensions, and to ensure that all human beings can fulfil their potential in dignity and equality and in a healthy environment.

Planet – this area about protection the planet from degradation, including through sustainable consumption and production, sustainably managing its natural resources and taking urgent action on climate change, so that it can support the needs of the present and future generations.

Prosperity – is determined to ensure that all human beings can enjoy prosperous and fulfilling lives and that economic, social and technological progress occurs in harmony with nature.

Peace – this is about being foster peaceful, just and inclusive societies which are free from fear and violence. There can be no sustainable development without peace and no peace without sustainable development.

Partnership – to mobilize the means required to implement this Agenda through a revitalized Global Partnership for Sustainable Development, based on a spirit of strengthened global solidarity, focused in particular on the needs of the poorest and most vulnerable and with the participation of all countries, all stakeholders and all people.

Actually, mentioned 5Ps of sustainable development could be applied for adopting traditional theories of economic growth, and this is partly a point of the latest research of Nobel prize nominees. The one of the well-known models of economic growth by R Solow based on ideas that growth is the function of capital (K) and labor (L): $\mathbf{Y = F(K, L)}$. As any model these one also simplified but it enables us to investigate how the main factors of production – labor, capital, technological change – affect the dynamics of production, when the economic system is in equilibrium state. The advantage of the Solow model is the delimitation of these factors and the gradual study of the impact of each of them on the process of long-term growth of national income. Further development of this model explain that labor should also be considered as a function of education (e): $\mathbf{L = F(e, L)}$, and also depends on the new way of using components of capital and labor – "innovation" (A) of the use of existing production factors: $\mathbf{Y = A * F(K, eL)}$. Underline that component A is not the only traditional vision of innovation in could and should include managerial, environmental, universal design issues.

The main idea of modern growth models is that the process is not limited solely by factors of production and their combination, but is a more complex and multidimensional process (Figure 3.1). Environmental and social issues became the inalienable part of modern models of economic development that could not be ignored. At present, the model of economic growth could look like $\mathbf{F = A * En(K, eL, N)}$, where, A – factor that determines the innovational part as a new use of all factors of production and their combinations; En – environmental aspects of development; K – physical capital; eL – human capital (including educational components); N – natural resources.

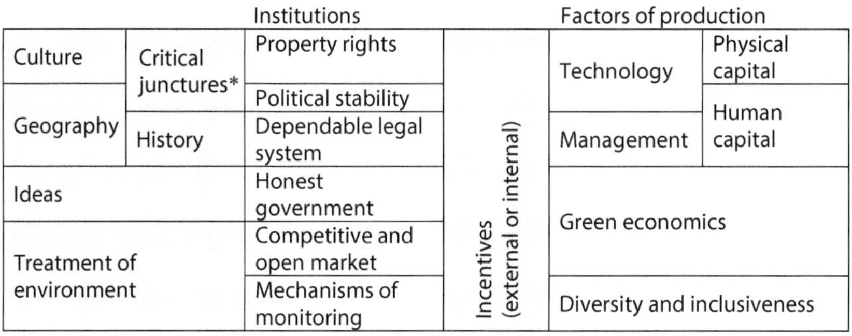

		Institutions		Factors of production	
Culture	Critical junctures*	Property rights		Technology	Physical capital
Geography	History	Political stability		Management	Human capital
		Dependable legal system			
Ideas		Honest government		Green economics	
Treatment of environment		Competitive and open market			
		Mechanisms of monitoring		Diversity and inclusiveness	

(Incentives (external or internal))

Figure 3.1 Factors of economic growth according to Solow model in author interpretation.

*Critical junctures are major events that disrupt the existing political and economic balance in one or many societies. Why Nations Fail (2012).

Nevertheless, we are more interested not in growth itself (like increasing in real national income/national output) but in development (as an improvement in the quality of life and living standards, e.g. measures of literacy, life expectancy, and healthcare). Ceteris paribus, we would expect economic growth to enable more economic development, e.g. higher real GDP enables more to be spent on health care and education. However, the link is not guaranteed but expected and these expectations reflected in indexes and rankings. The Global Competitiveness Index methodology show us 12 main pillars, which are institutions, infrastructure, macroeconomic environment, health and primary education, higher education and training, goods market efficiency, labor market efficiency, financial market development, technological readiness, market size, business sophistication, innovation. The Global Competitiveness Report 2017–2018 (2018). They perfectly suit our interpretation of modern economic growth models.

Actually, the work of the Nobel nominee 2018, Romer P. emphasized the necessity and effectiveness of the proactive approach to innovation – diversify and change the technologies and approaches before the effect of diminishing returns will slow down the growth and next step will require sufficient capital investments. This is about this innovative component – that can give us a clue how we can reach a success and why fest growth countries cannot overcome developed one for a long time. And this proactive position it is also about including environmental and social components in models of economic growth, that is mean that sustainable development began a ground for new models that should switch the attention from growth to development. Thus, the research of potential for economic development could be in the framework of social or environmental issues. In this paper we more concentrated on social issues, which could be concretized within the sustainable development goals Agenda 2030 (2015). We choose Goal 1: No Poverty, Goal 2: Zero Hunger, Goal 3: Good Health and Well-being, Goal 4: Quality Education, Goal 5: Gender Equality, Goal 10: Reduced Inequality, Goal 17: Partnerships to achieve the Goal. All these goals could be gathered together under the umbrella of equality issues.

Equality is ensuring individuals or groups of individuals are not treated differently or less favourably, on the basis of their specific protected characteristic, including areas of race, gender, disability, religion or belief, sexual orientation and age. Promoting equality should remove discrimination in all of the aforementioned areas. Bullying, harassment or victimization are also considered as equality and diversity issues. Understanding equality (2018).

Generally, we are talking about social equality as a state of affairs in which all people within a specific society or isolated group have the same status in certain respects, including civil rights, freedom of speech, property rights and equal access to certain social goods and services. The standard of equality that states everyone is created equal at birth is called ontological equality Benhabib (2002). There are several approaches to understand it:

- Equality of opportunity – the idea that everyone has an equal chance to achieve wealth, social prestige, and power because the rules of the game, so to speak, are the same for everyone.
- Equality of condition – the idea that everyone should have an equal starting point to create fairer competition in society. Here is where social engineering comes into play where we change society in order to give an equality of condition to everyone based on race, gender, class, religion etc. when it is made justifiable that the proponents of the society makes it unfair for them.
- Freedom of individuals – in order to have individual freedom there needs to be equality of condition which requires much more than the elimination of legal barriers: it requires the creation of a level playing field that eliminates structural barriers to opportunity.
- Equality of outcome – a position that argues each player must end up with the same amount regardless of the fairness (nobody will earn more power, prestige, and wealth by working harder).

Important to understand what type of equality is prioritized at the moment because it gives understanding of what should be done. According to survey of the Global Shapers Community (2017), 51% of people under 30 believe 'equal access to opportunities for all' is the most important thing for a free society – even more than job security.

3.2 Contemporary Peculiarities of Economic Development

The first thing that could not be ignored is growing. Latest studies about world growth demonstrate the progress and perspectives for almost all countries (Figure 3.2). The recent acceleration in world gross product growth stems predominantly from firmer growth in several developed economies, although East and South Asia remain the world's most dynamic regions. Cyclical improvements in Argentina, Brazil, Nigeria and the Russian Federation, as these economies emerge from recession, also explain roughly a third

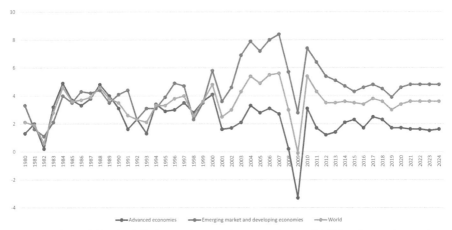

Figure 3.2   Real GDP growth, annual percent change (IMF DataMapper).

of the rise in the rate of global growth between 2016 and 2017 (World Economic, 2018). But recent economic gains remain unevenly distributed across countries and regions, and many parts of the world have yet to regain a healthy rate of growth. Economic prospects for many commodity exporters remain challenging, underscoring the vulnerability to boom and bust cycles in countries that are overly reliant on a small number of natural resources. Moreover, the longer-term potential of the global economy carries a scar from the extended period of weak investment and low productivity growth that followed the global financial crisis.

The Economist Intelligence Unit expects that a new generation of local companies will be riding this wave, determined not only to benefit from growth in their domestic markets, but also to expand internationally Going global (2018). By 2022 non-OECD countries will account for over 80% of the global population, and 40% of global GDP in nominal US dollar terms. Data indicates that real growth in consumer expenditure in non-OECD countries will average 4.8% per year in 2018–22, compared with 1.8% per year for OECD countries (Figure 3.3).

Next thing – is urbanization. It needs to underline that today only 600 urban centers generate about 60 percent of global GDP. While 600 cities will continue to account for the same share of global GDP in 2025, this group of 600 will have a very different membership. Over the next 15 years, the center of gravity of the urban world will move south and, even more decisively, east

Private consumption

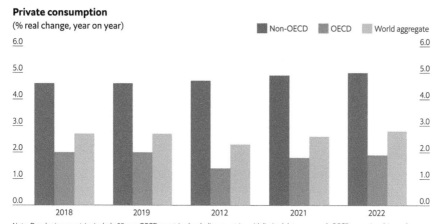

Note. Developing countries include 85 non-OECD countries (excluding countries with limited data coverage); OECD comprises 34 member countries; World data based on 120 countries (excluding countries with limited data coverage).

Source: The Economist Intelligence Unit.

Growth in developing countries, 2018-22

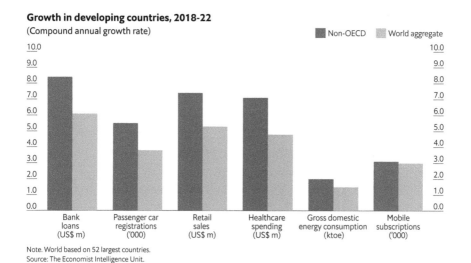

Note. World based on 52 largest countries.
Source: The Economist Intelligence Unit.

Figure 3.3 Characteristics of development of the world (Going Global, 2018).

(World Urbanization Prospects, 2018). This should change the pattern of the world with all inequalities and risk of living in the cities.

The variety of risks of economic development are much wider then economic problems and they are covering all aspects of life. Analysis of measuring risk to the global economy (Figure 3.4) illuminate that many potential

Global risk scenarios

■ Political   ■ Military   ■ Financial

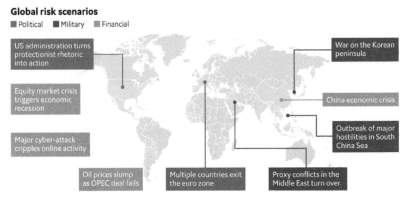

Figure 3.4   Global risk scenarios (Cause for concern? 2018).

challenges emerging from the world's largest economies. The Economist Intelligence Unit said that a number of risks, with their roots in the US, China and the EU. However, these risks are not limited to those geographies alone, and how they could morph into threats that destabilize large parts of the world, e.g.,

Prolonged fall in major stock markets destabilize the global economy;

Global trade slumps as US steps up protectionist policies;

Territorial disputes in the South China Sea lead to an outbreak of hostilities;

Global growth surges above 4%;

A major cyber-attack cripples corporate and government activity;

China suffers a disorderly and prolonged economic downturn;

There is a major military confrontation on the Korean Peninsula;

Proxy conflicts in the Middle East escalate into direct confrontations that cripple global energy markets;

Oil prices fall significantly after the Organization of the Petroleum Exporting Countries (OPEC) deal to curb production breaks down;

Multiple countries withdraw from the euro zone.

In addition, there are risks that either come from smaller regional hotspots, or are global in nature (Cause for concern? 2018).

Also, the changes in characters of work is should to be mentioned. The new work will (and actually have) a new structure. Today rather than moving up in one direction, ambitious employees will be able to move sideways, tapping into new networks. The survey The Future Work Place (2014) key findings was that in order to attract and retain high-calibre employees, companies need to foster a more collaborative environment. This might involve hot-desking, ideas workshops and regularly switching teams. Not only do employees respond well to this style of working, but corporations benefit too as it better equips them to compete with the startups that are disrupting their business.

Artificial intellect and robotization also should be taken into account. At this point we should understand that in midterm using AI will be complement to high-qualified workers and substitute to the low-qualified ones, but in long term period it will change the landscape of skills and qualifications of labor force and make labor market more flexible. The creating of so called "human clouds" (a vast global pool of freelancers who are available to work on demand from remote locations on a mind-boggling array of digital tasks – which is really set to shake up the world of work) also a result of technological developing and fasten spreading communicational services. The one of result of this are internet of things and internet of everything. Spreading of technologies gives an opportunity for development, education and working through the world without retirement and giving instruments for controlling and observation at the same time. The labor market endeavor to flexibility. By 2022, at least 54% of employees globally will require re- and up-skilling (The Future of Jobs Report, 2018). That's mean that the only support people in getting the training they need for jobs in the next five years is not enough, but we need to prepare young students with the skills to adapt to the types of jobs we will need in the next 20 years.

Summarizing all above few ways can be highlighted for mitigating challenges and building the successful strategy of development: invest in strengthening local and regional economies, implement and develop innovate educational institutions and aggressively close the skills gap, focus on the most vulnerable populations, stop climate change, build a movement focused on equity. Advancing the priorities above and creating greater equity will require a more coordinated global movement than exists today. Many businesses, NGOs, advocacy groups, academics and even individuals have unprecedented global reach and ability to influence equitable outcomes. Millennials are likely to reward businesses who participate in this movement, preferring to work at and purchase from businesses that are driving social good (The Deloitte Millennial Survey, 2018).

3.3 Economic Effects of Inequality – Finding the Balance Between Effectiveness and Efficiency

According to the one of 10 principles of economics that generally know as "people face tradeoffs" we need to find balance between effectiveness and equality of development keeping in mind difference between effectiveness and efficiency. While efficiency refers to how well something is done, effectiveness refers to how useful something is (Figure 3.5). Efficiency means society gets the most that it can from its scarce resources and in this framework, equity means the benefits of those resources are distributed fairly among the members of society. That is why in our research we concentrated on humans (members of society) not on inequality displays like income distribution, poverty, education etc. Thus, contemporary economics should not operate the only effectiveness as a main characteristic of successful strategy, but also efficient and equality.

We would like to focus our attention on what the present situation is. If we started with the macro level, we should mention that generally macroeconomic politics perceived like neutral in relation to the race, gender, ability, until the beginning analysis of its results. Macroeconomic politics in a different way affects markets of paid and unpaid work, e.g. changes in the budget that leads to cutting of social spending and rising of unemployment can raise demand on unpaid work which traditionally made by women. Another aspect – the liberalization of trade could negative influence the local low-paying workers (where the share of women or colored men or immigrants are bigger). The consequences of crises like cutting social expenditures and

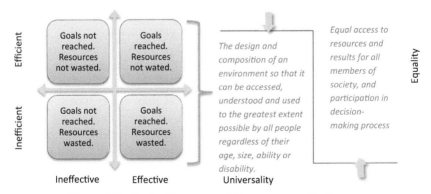

Figure 3.5   Goal achievement: efficiency versus effectiveness.

rising prices also affects more families with children (Table 3.1) and social workers, whose salaries depending from governmental payments.

In terms of individuals, some key factors that are making a person more "at risk" of inequality are:

- Unemployment or having a poor quality (i.e. low paid or precarious) job;
- Low levels of education and skills because this limits people's ability to access decent jobs to develop themselves and participate fully in society;
- The size and type of family;
- Gender – women are generally at higher risk of poverty than men as they are less likely to be in paid employment, tend to have lower pensions, are more involved in unpaid caring responsibilities and when they are in work, are frequently paid less even for the same job;
- Disability or ill-health because this limits ability to access employment and also leads to increased day to day costs;
- Being a member of minority ethnic groups and immigrants/undocumented migrants;
- Living in a remote or very disadvantaged community where access to services is worse.

The inequality is not reflected in macroeconomic indexes as well as negative environmental impact. This problem widely discussed in economic society and the latest research of William D. Nordhaus gives the methodology for integrating climate change into long-run macroeconomic analysis (Facts, 2018); but there is still no solution for including equality and gender in macroeconomic indexes and policies.

Nevertheless, data prove that economies with more equality perform better and associate with the developed economies (Figure 3.6). This fact commonly interpreted like equality is intrinsic for stile of life of developed countries, but better interpret this fact that achieving the equality is a step towards the developing. In this case it gives us a clue not a challenge. The research of McKinsey Global Institute (2015) said that if every country matched the progress toward gender parity of its fasters-improving neighbor, global GDP could increase by up to $12 trillion in 2025.

All mentioned negative aspects of inequality changes affected mood of people, their loyalty and willingness to participate in social life. That in general determines the effectiveness of the state policy. From this perspective tracking such indexes like relative poverty can give information for adjusting the strategy towards equality. Relative poverty is where some people's way of life and income is so much worse than the general standard of living in the

Table 3.1   Severe material deprivation* by household type, 2017 (early data) – % of population

Country	Single Female	Single Male	Single Person With Dependent Children	Two Adults with One Dependent Child	Two Adults with Three or More Dependent Children	Two Adults At Least Aged 65 Years or Over
Belgium	8.6	9.6	16.7	3.9	4.2	1.7
Bulgaria	52.6	37.1	49.0	20.8	77.7	30.4
Czech Republic	7.0	7.5	11.6	3.0	5.7	2.0
Denmark	4.5	8.0	7.7	1.9	1.8	0.2
Germany	8.8	7.6	10.3	0.9	4.5	0.7
Estonia	9.9	9.3	9.2	2.9	3.4	3.0
Greece	21.7	23.2	36.1	17.2	29.0	13.4
Spain	4.5	7.3	11.8	4.2	9.2	2.3
France	6.9	6.0	10.6	2.1	5.0	1.1
Croatia	19.6	24.8	20.1	7.6	10.2	13.7
Italy	11.1	12.4	12.8	7.0	10.2	7.1
Cyprus	8.1	14.9	22.8	11.4	15.7	4.9
Latvia	18.1	18.8	18.1	6.3	11.2	12.1
Lithuania	21.7	14.8	25.9	7.2	20.8	11.8
Hungary	17.2	19.5	29.9	12.3	25.1	8.9
Malta	4.4	7.7	20.8	1.8	3.6	1.9
Netherlands	5.5	7.7	9.5	1.5	2.1	0.6
Portugal	11.5	12.1	17.0	4.5	14.2	6.6
Romania	27.6	25.6	23.1	11.1	28.9	18.2
Slovenia	11.7	11.9	11.8	3.3	1.6	4.0
Finland	4.5	5.6	8.6	0.5	1.1	0.5
UK	5.5	7.7	20.0	3.8	8.6	0.7
Norway	2.7	6.0	8.8	0.3	2.4	0.1

*Deprivation indicators is an approach to measuring relative poverty. These are an attempt to move beyond just monetary indicators and to take better into account the actual standard of living that people enjoy. Essentially the approach involves identifying goods or activities which are seen as basic necessities in the country someone is living. In some country's poverty is measured by combining relative income lines with deprivation indicators.
Source: Eurostat.

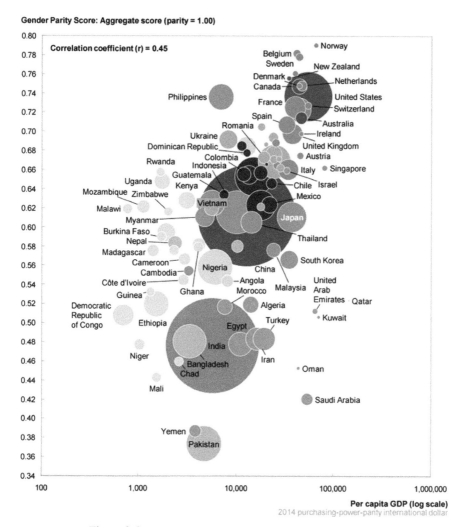

Figure 3.6 Interconnection between gender parity and GDP.

Source: https://ourworldindata.org and Esteban Ortiz-Ospina and Max Roser (2018).

country or region in which they live that they struggle to live a normal life and to participate in ordinary economic, social and cultural activities. What this means will vary from country to country, depending on the standard of living enjoyed by the majority (The measurement, 2011). In 2010, the EU adopted the Europe 2020 Strategy, aimed at guiding the EU towards a smart sustainable and inclusive economy.

Another aspect why equality matters for economic policy of the country is the differences in spending structure according to the gender. Women's income is more likely to be reinvested in the family and community where they live, which contributes to their economic development (in particular education, nutrition and health). A study conducted in Brazil has shown that the chance for child's survival has increased by 20% when the mother has the ability to dispose of finances (OECD, 2011). However, the restrictions on the availability of and access to finance for women are often justified by cultural traditions, health status, social hierarchy, etc.

In the US, every $1 that the government invests in family planning saves $7 taxpayers, and in developing countries, such as Jordan, up to $16 (Family Planning, 2009). The Copenhagen Consensus Center has illustrated that every dollar spent on modern contraceptive methods will give $120 in common benefit (Benefits and Costs, 2015). Investing in educational meetings with families and integrating gender perspectives into corporate culture is also reflected through increased productivity, shorter workouts, greater focus on results etc. The study of the effect of introducing elementary healthcare opportunities for women with children in Bangladesh (USAID, 2007) and Egypt HERproject (2011) had investment return of $3: 17 and $4: 17, respectively.

Control over income is a step to overcome inequality, and not vice versa. Studies show that women's ability to earn and participate in the distribution of family budgets will increase the economic well-being of the family. The Bolsa Familia program (2016) in Brazil, which provides social transfers directly to the wives, recorded a reduction of 25% inequality and 16% – families living in extreme poverty.

The combined impact of growing gender parity, a new middle class in emerging markets and women's spending priorities is expected to lead to rising household savings rates and shifting spending patterns, affecting sectors such as food, healthcare, education, childcare, apparel, consumer durables and financial services (Goldman Sachs, 2009). With women controlling 65% of global household spending and estimated global consumer spending of currently US$40 trillion there are large potential benefits for companies with employees who can understand diverse customer bases (Buying Power, 2015).

Research of inequality on the microlevel give us more precise gaps in modern management and leadership. Based on four years of data from 462 companies employing more than 19.6 million people, including the 279 companies participating in this year's study, two things are

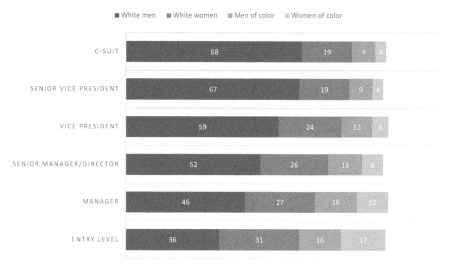

Figure 3.7 Representation by corporate role, by gender and race in 2018, % of employees.
Source: Women in the Workplace Study (2018).

clear: one, women remain underrepresented, particularly women of color (Figure 3.7).

That's mean that generally managerial systems, approaches and practices were built by white men for white men and other participants discriminated and treated as a labor force in ineffective and inefficient way. When women reach the upper ranks of power, they're put into precarious positions, meaning there's a greater risk for them to fall. There is observed effect of glass cliff – a phenomenon whereby women (and other minority groups) are more likely to occupy positions of leadership that are risky and precarious Ryan and Haslam (2005). According to the research during periods of overall stock market decline, firms that brought women to their boards were likelier to have experienced a consistently bad performance in the preceding five months than those who brought on men. That's mean then you move somebody from minorities to the leading position according to the qualifications not superstitions you use the potential more effective and easily find the better solutions. From other point of view, the only chance for discriminated part of labor force to take the position is to take risks that in the "traditional" managers and CEOs do not what to face taking care of their reputation.

Although the representation of women in politics has increased – over the past 20 years, three quarters of the total number of women in these positions have been elected or appointed to presidential and prime ministerial positions.

However, the overall gap between men in the realization of political rights and opportunities for women is 23%. Only 20% of all national parliamentarians are women, 18% are ministers and only 47% of countries have head of state – women in the last 50 years (Gender Gap Report, 2017).

3.4 Conclusions: What Can Be Done?

The equality is the bases for economic growth regardless of level of economic development and prosperity of the country. The inequality has a lot of issues, participants and manifestations, that could change in according with knowledge development, innovations and technologies spreading. The equality should not be treated like a panacea from poverty and instability. Inequality could not be excused by the recession, crises, traditions etc. The equality must become the clue for economic growth and development for the number of reasons:

- Sustainable macroeconomic development that could be reached also by including equality issues in agenda. It gives possibility to build up more balanced strategy and effective governance, mitigate signs of crises.
- Effectiveness of using the resources. If think about three main factors of production (natural resources, capital and labor) as a basis for economic development we should strive to their effective using and allocation. From this prospective discrimination and inequality narrowed our possibilities for development through not using all possible and available factors of production.
- Rising the competitiveness of the business. The discrimination negatively affects businesses through low job satisfaction and motivation, commitments before other worker and a firm, decreasing of loyalty and rising the turnover, spoiled reputation and problems with recruitment, low liability. Thus, equality and transparency inside the organization makes it more attractive and more competitive on the market. Moreover, in todays connection age the reputation of the firm forming by workers and former workers that could be like advisers, significant unformal advertisers and consumer of product and services.
- Increasing the overall level of income and reduce the income gap by creating a more flexible regime of work in most areas of economic activities. This allows to attract more workers and more efficient use their potential and at the same time give them more time and opportunities to spend their free time and money for reinvesting them into the economy.

– Reshape the human capital by widening the access to all attainments of society – intellectual, technological, materialistic, cultural etc. This is in long perspective will create the basis for "innovative" component of economic development which is critically important for continued development.

To close the gender gap and overcome inequality we need to change how we think. Today a study published by the Harvard Business Review (2011), that asked college students to identify whether a man or woman should be chosen to head a theoretical organic food company after a CEO retired, found that when the company had been led by a man and was doing well, 62% of students chose a candidate who was a man. When the man-led company was in crisis, 69% said a woman candidate would be better.

Likewise, a study by the National Bureau of Economic Research, which argues that young women are inclined to restrain their career aspirations in men surrounded if they are not married or have long-term relationships (Bursztyny et al., 2017). Researchers have come to the conclusion that even in highly educated and educated environments, men are more committed to women-oriented homes and this holds back women and forces them to choose less-paid work with a flexible schedule. According to the poll, almost 75% of single women abandoned career growth so as not to create an overly persistent and arrogant image against 60% of women in relationships, 50% of men in relationships and 43% of single men.

To overcome this situation in the future we should change the way of thinking today. The perception of equality should not be like a challenge but as a chance make it incentive to a chain of changes. For example, implementing transparent procedures for applying to the position in the firm including salary creates not only the basis for fair competition but makes the financial system clearer, and simplify inside and outside communications. That's why the key for all these changes are education and raising the awareness. The people could not implement any idea if they do not know about it and do not understand what it for. The next – is implementation on all levels. Of course, policies of equality in the international or multinational corporations are more visible, but it does not mean that a number of small initiatives are less effective. The empowerment of women during the crises (glass cliff phenomena) gives a chance to change the situation just by fixing the new status quo for longer time for breaking the stereotypes.

The main principle is "little strokes fell great oaks" – every initiative is important and should not be ignored or neglected. And issues of inequality and discrimination are urgent for all types of societies and economies, and every country could build up its own Agenda on the basis on equality as a resource for long-term stable and sustainable development.

Acknowledgment

I'd like to express gratitude to initiatives "Gender in details" https://gender indetail.org.ua/) and "Ukrainian women in UN" (https://www.facebook.com/ UkrainianWomenUN/) for inspiring and empowerment during my work.

References

Agenda 2030 (2015). For Sustainable Development: Transforming our World – 17 session of UN General Assembly A/RES/70/1 – https://www.unfpa.org/resources/transforming-our-world-2030-agenda-sustainable-development.

Benefits and Costs (2015). Of the Population and Demography Targets for the Post-2015 Development Agenda. Copenhagen Consensus Center – http://www.copenhagenconsensus.com/sites/default/files/population_assessment_-_kohler_behrman_0.pdf.

Benhabib, S. (2002). The claims of culture: Equality and diversity in the global era. Princeton University Press.

Bursztyny, L., Fujiwara, T. and Pallai, A. (2017). Acting Wife: Marriage Market Incentives and Labor Market Investments – http://scholar.harvard.edu/files/pallais/files/acting_wife.pdf.

Buying Power (2015). Global Women, Catalyst – www.catalyst.org/knowledge/buying-power-global-women.

Cause for concern? (2018). The top 10 risks to the global economy – A report by The Economist Intelligence Unit – 25 p.

Esteban Ortiz-Ospina and Max Roser (2018). Economic inequality by gender. Published online at OurWorldInData.org. Retrieved from: https://ourworldindata.org/economic-inequality-by-gender [Online Resource].

Facts (2018). William D. Nordhaus. NobelPrize.org. Nobel Media AB 2018. Tue. 6 Nov 2018. – https://www.nobelprize.org/prizes/economic-sciences/2018/nordhaus/facts/.

Family Planning and the MDGs (2009). USAID Health Care Programme – http://www.healthpolicyinitiative.com/Publications/Documents/788_1_Family_Planning_and_the_MDGs_FINAL_June_09_acc.pdf.

Gender Gap Report (2017). World Economic Forum – http://reports.weforum.org/global-gender-gap-report-2017/.

Going Global (2018). Key corporate trends in developing markets // A report by The Economist Intelligence Unit.

Goldman Sachs (2009). Global Markets Institute, The Power of the Purse: Gender Equality and Middle-Class Spending.

Harvard Business Review (2011). Susanne Bruckmller, Nyla R. Branscombe. How Women End Up on the "Glass Cliff" – https://hbr.org/2011/01/how-women-end-up-on-the-glass-cliff.

HERproject (2011). Health Enables Returns. Racheal Yeager, BSR – https://www.bsr.org/reports/HERproject_Health_Enables_Returns_The_Business_Returns_from_Womens_Health_Programs_081511.pdf.

Human Rights and Poverty Reduction (2004). A Conceptual Framework. United Nations New York and Geneva – https://www.ohchr.org/Documents/Publications/PovertyReductionen.pdf.

IMF DataMapper – https://www.imf.org/external/datamapper/NGDP_RPCH@WEO/OEMDC/ADVEC/WEOWORLD.

McKinsey Global Institute (2015). Report the power of parity: How advancing women's equality can add $12 trillion to global growth – https://www.mckinsey.com/featured-insights/employment-and-growth/how-advancing-womens-equality-can-add-12-trillion-to-global-growth.

OECD (2011). Women's economic empowerment – http://www.oecd.org/dac/gender-development/womenseconomicempowerment.htm.

Our Common Future (1987). Report of the World Commission on Environment and Development – World Commission on Environment and Development A/42/427 – http://www.un-documents.net/our-common-future.pdf.

Ryan, M. K. and Haslam, S. A. (2005). The Glass Cliff: Evidence that Women are Over-Represented in Precarious Leadership Positions – British Journal of Management, Vol. 16, 81–90. https://is.muni.cz/el/1423/jaro2017/VPL457/um/62145647/Ryan_Haslam_The_Glass_cliff.pdf.

SMD (2018). Severe material deprivation rate – EU Open Data Portal – https://data.europa.eu/euodp/en/data/dataset/MTKIPP1CLPtgYROgthUAGw.

The Bolsa Familia program (2016). Women Deliver – https://womendeliver.org/2016/case-study-brazils-bolsa-familia/.

The Deloitte Millennial Survey (2018). Millennials' confidence in business, loyalty to employers deteriorate. Deloitte Insights – https://www2.deloitte.com/global/en/pages/about-deloitte/articles/millennialsurvey.html.

The Future of Jobs Report (2018). World Economic Forum – https://www. weforum.org/reports/the-future-of-jobs-report-2018.

The Future Work Place (2014). Key trends that will affect employee well-being and how to prepare for them today. Unum Limited. – https:// www.unum.co.uk/hr/the-future-workplace.

The Global Competitiveness Report, (2017–2018). World Economic Forum – http://reports.weforum.org/global-competitiveness-index- 2017-2018/ #topic=about.

The Global Shapers Survey (2017). Annual Survey 2017. #ShapersSurvey – http://shaperssurvey2017.org/.

The measurement of extreme poverty in the European Union (2011). European Commission. Directorate-General for Employment, Social Affairs and Inclusion – http://ec.europa.eu/social/BlobServlet?docId=6462&l angId=en.

Understanding equality (2018). Equality and Human Rights Commision – https://www.equalityhumanrights.com/en/secondary-education-resourc es/useful-information/understanding-equality.

USAID (2007). Effects of a workplace health program on absenteeism, turnover, and worker attitudes in a Bangladesh garment factory – http:// pdf.usaid.gov/pdf_docs/pnaec188.pdf.

Why Nations Fail (2012). The Origins of Power, Prosperity, and Poverty. Aaron Acemoglu and James A. Robinson – NY, 462 p.

Women in the Workplace Study (2018). Study by Alexis Krivkovich, Marie-Claude Nadeau, Kelsey Robinson, Nicole Robinson, Irina Starikova, and Lareina Yee. Leanin.Org and McKinsey – https://www.mckinsey.com/ featured-insights/gender-equality/women-in-the-workplace-2018.

World Economic. Situation and Prospects (2018). United Nations: New York.

World Urbanization Prospects (2018). United Nations. DESA/Population Devision – https://population.un.org/wup/.

4

Green Investment as An Economic Instrument to Achieve SDGs

Olena Chygryn, Tetyana Pimonenko* and Oleksii Lyulyov

Marketing Department, Sumy State University, Ukraine
E-mail: tetyana.pimonenko@gmail.com
*Corresponding Author

4.1 Introduction

The ongoing snowballing economic, social and ecological development from one side provokes the improving of the countries' welfare, from the other side lead to increase of the negative anthropogenic effect on the environment. The world community has already made a lot in that direction. Thus, a lot of regulations and stimulation laws, action plans and agendas, protocols, instruments have been accepted and implemented. Thus, the latest document was 'The 2030 Agenda for Sustainable Development' (The 2030, 2018) which indicated 17 sustainable development goals for achieving in 2030, where highlighted the importance of renewable energy, as it will support the achievement of climate policy goals. Noted, that the characteristics of renewable energy investment have received greater attention from energy policymakers and academics. In 2015, the share of renewables in total electricity generation was 23%, while a record amount of 285.9 billion US dollars was invested in renewable (IEA, 2016). Therefore, all abovementioned problems and goals require putting a powerful investment for their solving and achieving.

It should be underlined, that it is not so big issue for the developed countries, but a lack of financial resources is a huge issue for the developing countries (Chygryn, 2018). On the other side, investment in green projects and activities are not attractive to investors. Firstly, it is the consequences of the existing stereotypes that investment in greening is non-profitable.

Secondly, such investments have a huge payback period which negative influence on making the decision if invest or not. Thus, the increasing number of ecological problems contribute to accept prompt actions for developing and implementing the corresponding instruments, which could resolve abovementioned issues and not retain the economic development in the country. Moreover, these actions should consider the current modern trends in the world economy. Lastly, it is necessary to indicate and summarize the benefits and positive effects of instruments from the different point of views for all stakeholders with the purpose to promote and popularize such instruments.

Despite ample global savings and record-low long-term interest rates, infrastructure investments in sustainable development projects are often unable to attract long-term private financing, and the costs of financing are relatively high – in some cases prohibitively so. While the volume of private finance including cross-border finance has grown rapidly over the past two decades, very little of this capital is being directed toward long-term investment, and even less is being made available for infrastructure financing. Improving access to and reducing the cost of private capital for sustainable infrastructure will require concurrent actions on several fronts: deepen domestic capital markets, enhance and scale up risk mitigation instruments, develop infrastructure as an asset class, expand the range of financial instruments, greening of the financial system. In this direction, the traditional economic and ecological instruments should be adopted and modernised accordingly to the ongoing market economy. Thus, a very promising instrument is the green investment, which integrated the main aspects of the traditional investments and features of a green economy.

4.2 Method

Under this investigation the authors used the traditional and modern methods of scientific knowledge: analysis and synthesis – in identifying trends of green investment in the developed countries; comparison and compilation – to analyse the experience of developed countries to support and develop green investment with purpose to allocate resources for achieving the Sustainable Developments Goals 2030; the statistical and mathematical methods – in analysis of the economic, social and ecological benefits of green investment for stakeholders; the scientific support methods – to summarise and to formulate conclusions on perspectives to develop green investment in Ukraine. These approaches allow allocating the challenges and opportunities for Ukraine to develop green investment as additional financing for achieving the Sustainable Development Goals 2030. In addition, it allows taking into account the best EU practice on supporting green investment in Ukraine conditions. The main purpose of this paper is to analyse the potential of green investment in the EU and Ukraine. In this case, it is necessary to highlight the ecological, economic and social benefits of green investment for stakeholders.

4.3 Results

The results of the analysis showed that green investment is a complex definition, which involved green and economic aspects. Noticed, that 'green' is the general and broad term which defined by the scientists from the different points of views: philosophy, social, technical and economic. The experts in the work "Defining and Measuring Green Investments" indicated that definitions of "green" can be explicit or implicit. Some are very broad and generic; others are more technical and specific. Some are investment-driven; others come out of ecological or ethical discussions (Inderst et al., 2012).

On the other side, in general, investment is time, energy, assets; resources were spent with the purpose to receive future benefits. From the economic point of view, investment is the purchase of goods that are not consumed today but are used in the future to create wealth. Most spread definition in finance sense, investment is a monetary asset purchased with the idea that the asset will provide income in the future or will later be sold at a higher price for a profit (Investment, 2018). It should be noted, that among scientists there are some narrow interpretations of the concept "green investment" and the nature and coverage of a wide meaning. In addition, a mostly green investment associated with socially responsible investing, environmental, social and governance investing, sustainable, long-term investing or similar concepts (Inderst et al., 2012).

Traditionally, green investment is directed to manufacture which specializing on cleaning equipment and waste management, environmental control facilities and monitoring devices and etc. The authors Eyraud et al. (2013) and Martin and Moser (2016) defined green investment as the investment which direct to the decreasing of CO_2 emission. Noted, that in the paper (Adeel-Farooq et al., 2018) the authors defined green investment as greenfield investment and associated it with the capital which finances the green growth. The group of the scientists in the papers (Hagspiel et al., 2018; Cebula et al., 2015; Yevdokimov et al., 2018; Prokopenko et al., 2017) defined green investment as an investment in renewable energy. Mielke and Steudle (2018) analysed green investment as funding in technologies and projects for climate change mitigation (Pimonenko et al., 2017).

The experts of Triodos Bank identified green investment as the financial products which guarantee not only financial benefits but also environmental and social effects (Triodos, 2018). During Stream "Our green future: green investment and growing our natural assets" the scientists underlined that green investing is the concept which involves social and environmentally responsible corporate governance (Djalilov et al., 2015; Chigrin, 2014). In addition, they highlighted that green investment is a way to decrease invest-ment risk and at the same time, to promote green development (Summary, 2018).

The obtained results of analysis of approaches to define green investment among Ukrainian scientists showed that green investment defined as a capital which finance green projects (Andreeva, 2015; Vasylyeva et al., 2014), eco-nomic activities for developing ecosystems (Arestov, 2015) and for achieving of green and sustainable growth (Vyshnitskaya, 2013; Kvaktun, 2014). The consolidated approaches to defining terms green investment are presented in Table 4.1.

Thus, the green investment could be defined as a monetary asset pur-chased on green goals with the purpose to achieve income and positive green effect in the future for achieving sustainable and green growth. In this case, green goals are: mitigate climate change, develop alternative energy resources, develop clean technology and etc.

Accordingly, the complexity of the term green investment determines the following features:

- Orientation to use, protection, reproduction conditions in support of natural resources;
- The investee has the general public character for many consumers and users and often the problem cannot be solved by a separate subject, region, country;
- The need to consider different sources of investment, combined in time and space, their forms and types;
- Natural system (assimilations potential), its elements cannot be dis-counted, although they may reduce, lose original properties under the influence of anthropogenic factors;
- Differ from the investment that provides the state, interstate, own, mixed forms of organization of social and economic activities of natural users;
- Considering specific properties of self-regulation and heal itself ecosys-tems its individual components (Krasnyak et al., 2015).

Table 4.1 Approaches to define "green investment"*

Main Goal	Contents
Prevention and liquidation pollution	Green investment is all types of property and intellectual values invested in order to mitigate climate changes and decrease the CO_2 (Andreeva, 2005; Eyraud et al., 2013; Mielke and Steudle, 2018).
The development of ecosystems	Green investment is not only environmental investments, and any investment aimed at developing ecosystems (Arestov, 2010).
The social and ecological and economic effect	Eco investment is the all types of property and intellectual values which invested in economic activities with goals: to reduce negative anthropogenic impact on the environment, eco-destructive impact of the product and service during life circle; conservation, effective using of natural resources and improving natural resources areas; to ensure environmental security of the country (Vyshnitskaya, 2013).
The implementation of environmental protection measures	Green investment is funds investing only into environmental protection measures (Anischenko, 2007).
Creating special funds	Green investment is capital that is aimed to develop of profitable assets in the production and exploitation which, firstly, lead to reduce the utilization of natural resources and, secondly, to soften (or liquidated) negative impact on the environment and human health. (Kvaktun, 2014; Chygryn, 2016).
Green growth	Green investment is an investment in green and sustainable growth (The Green, 2013; Adeel-Farooq et al., 2018).
Investment in green projects and clean technology	Green investment is essential investment activities that focus on companies or projects that are committed to the conservation of natural resources, the production and discovery of alternative energy sources, the implementation of clean air and water projects, and/or other environmentally conscious business practices (Investopedia, 2018; Martin and Moser, 2016; Hagspiel et al., 2018).
Environmental and social performance	Green investment is the financial products that take into consideration issues wider than purely financial performance, such as environmental and social concerns (Triodos, 2018).
Promoting green development	Green investment is a concept involves socially and environmentally responsible corporate governance as a way to reduce investment risk and at the same time, to promote green development (Summary, 2018).
Improving the environment	Green investment is traditional investment instruments (such as stocks, exchange-traded funds and mutual funds) in which the underlying business(es) are somehow involved in operations aimed at improving the environment. This can range from companies that are developing alternative energy technology to companies that have the best environmental practices (Green, 2018).

*Compiled by the authors.

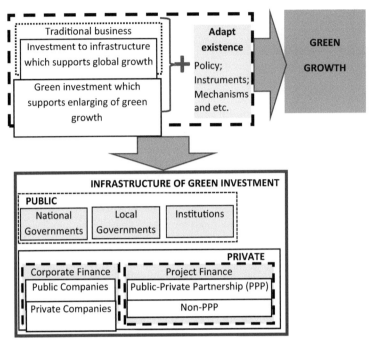

Figure 4.1 Principal scheme of transition from the existing model of business to green growth*.

*Developed by the authors.

It should be underlined, that additional costs of greening growth are insignificant compared with the costs of fuel savings compensating in large part for the investment requirements. Nevertheless, there are a huge number of barriers which should be overcome through the reorientation the mind of the business vision from the overusing to green.

In addition, a lot of existing policies, instruments and mechanisms should be modernized and adapted accordingly to the new market economy. The principal scheme of transformation from traditional business to green growth is shown in Figure 4.1. It should be underlined, that green investment has also own infrastructure, sources and channels (Figure 4.1).

Green investment has distinctive characteristics compared to investments in other economic spheres. It is due to the fact that green investment is not directed to a profit. Although in the long-term appropriate investment projects can be profitable or create conditions for a high level of profitability of other investment projects. Thus, worldwide experience often shows the absolute economic efficiency of investments. As modern innovative production and business activities are impossible without saving technologies and monitoring of environmental cleanliness products. Along with it, an important role in boosting green investment belongs to the government which at the international, national, regional and local levels should develop appropriate conditions for promoting and encouraging the green investment. In this perspective, public administration, green investment should not be regarded only as investment in regeneration, protection and development of the environment, but also in creating conditions, motivation and encouragement to attract capital in the economic spheres, the reproduction and development of environmental consciousness of society as a whole (Chygryn, 2015).

The results of the analysis showed that the classification of green investment isn't so different from general investment. The scientists in the report (Inderst et al., 2012) highlighted, that green investment is closely connected with other types of investing and allocated the main approaches to classify green investment as follows:

– Green (eco-friendly, climate change, etc.) investing;
– The "E" in ESG (environmental, social and governance) investing;
– Thematic investing (in green sectors or themes such as water, agriculture);
– SRI (socially or sustainable responsible investing);
– RI (responsible investing);
– SI (sustainable investing), sustainable capitalism;
– Impact investing (including microfinance);
– Long-term investing;
– Universal ownership concept;
– Double or triple-bottom-line investing (with financial, social and ecological goals) (Inderst et al., 2012).

On the other side, the authors in their papers (Vyshnickaya, 2013; Kvaktun, 2014; Heinkel, 2003) highlighted that green investment involves all abovementioned investment, and these are types of green investment. However, the experts from the EU commission assumed, that green investment could be classified by the green assents. The authors in the paper (Ibragimov et al., 2019) proved that defining the meaning of green investment related to the assets type. In this case, the EU commission is going to develop the classification of the green assets with the purpose to avoid "greenwashing". The results of analysis allow systematising the main approaches to classify green investment by groups follows as the level of investment; stakeholders;

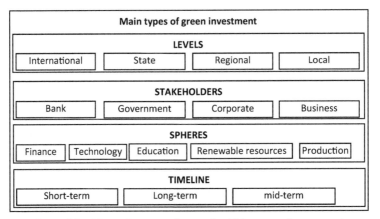

Figure 4.2   The main approaches to classifying green investment*.

*Developed by the authors.

spheres and timeframe. Each group involves different types of investments. The summarising results showed in Figure 4.2.

The experts in the paper (Bhattacharya et al., 2016) proved that developing green infrastructure require to attract additional finance recourses as the green investment. Thus, according to the Report "Better Finance, Better Development, Better Climate" the energy sector needs $40 trillion, transport $27 trillion, water supply and sanitation – $19 trillion. Furthermore, the authors highlighted the huge investment gap by sectors and level of countries development. The detail information on infrastructure needs presented in Table 4.2.

The results of the analysis showed that in the world practice green investment is the most popular in renewable resources and financial sectors. In addition, according to the survey of corporate pension funds reports (Eurosif, 2011) the equities and bonds are well ahead of other asset classes for green investment. The experts in the paper (Inderst et al., 2012) wrote that the same results were shown in another survey by (Novethic, 2011). It questioned 259 institutional investors of Europe with assets of EUR 4.5 tn. Thus, green investment is most popular among equities (40% fully implemented), followed by corporate bonds (31%), government bonds (24%), real estate (19%), private equity (15%), money market funds (14%) and commodities (8%) (Inderst et al., 2012). Therefore, green bonds market is increasing from year to year in the world (Figure 4.3). The similar conclusions made by the group of scientists in the paper (Chygryn et al., 2018).

Table 4.2 Estimation of green investment in infrastructure for SDG in the world, (2015–2030), USD trillions

Source	Energy	Transport	Water Supply & Sanitation	Telecom	Total
OECD (2006)	3.9	6	17	5.9	32.8
OECD (2012)		9.6			
Boston Consulting Group (2010)	4	7	14.3	9.2	34.5
McKinsey Global Institute (2013)	13.2	25.8	12.7	10.3	62
World Bank (2013)	4	4.1	3.2	2.8	14.1
International Energy Agency (2014)	37.6				
NCE, 2014 (Total Needs – BAU)	50.4	14.8	23.1	7.7	96.1
NCE, 2014 (Total Needs – Low Carbon)					101.6
NCE (Core Infrastructure BAU)	11	14.8	23.1	7.7	56.7
NCE (Core Infrastructure – Low Carbon)	−1.1				55.6
NCE (Energy – Primary Generation and Use)	39.4				39.4
UNCTAD (2014)	9.5–14.4	5.3–11.7	6.2		24.5–38.9
UN Sustainable Development Solutions Network (2015)	5.3–5.4	9.5	0.03		19.8–19.9
Brookings (2016)	20.5–23.9	27.2–31.4	12.7–14.7		74.7–86.6

Recourse: Bhattacharya et al., 2016.

The share of green bond in the debt market is 1% among G20 countries. However, the growth is significant: worldwide annual issuance rose from just US$ 3 billion in 2011 to US$ 95 billion in 2016 (Climate, 2017).

According to the official dataset in 2016 the biggest progress at the green bond market had the following countries: France, Germany, Mexico and South Africa. France issued the first green sovereign bond in early 2017 – the largest green bond issued to date at EUR 7 billion – increasing the overall French green bond market by approximately 25% (Climate, 2017).

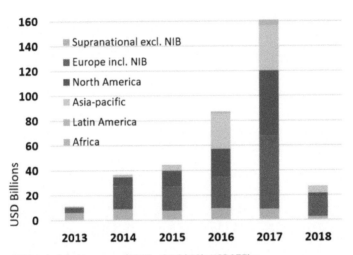

** Total global issuance (2007 - Q1 2018): USD377bn*
European issuance since 2010, i.e. first European issue: USD141bn

Figure 4.3   Dynamic of the green bond market in the world.

Source: The green bond, 2018.

China is just behind the EU with regard to market penetration. The first Chinese green bonds were issued in late 2015, but substantial growth had been made by China 2016 largest single green bond issuing country (Climate, 2017).

According to the official report (Green bonds, 2018) developed by Climate Bond Initiatives, the goal is US$1 trillion of annual green bond issuance by 2020. According to Bloomberg New Energy Finance estimation in 2017 investment in clean energy reached $333.5 billion, up 3% from a revised $324.6 billion in 2016, and only 7% short of the record figure of $360.3 billion, reached in 2015 (Bloomberg, 2018). In addition, the Chinese invested in 2017 to the clean energy more than $132.6 billion, up 24%, setting a new all-time record. The next biggest investing country was the U.S., at $56.9 billion, up 1% in 2016. In third place, Japan saw an investment decline by 16% in 2017, to $23.4 billion (Bloomberg, 2018).

The results of the analysis showed that among Asia Pacific (APAC) Europe, Middle-East and Africa (EMEA), North, Central and South America (AMER) regions the biggest sum of green investment in clean energy was in APAC region – $187 billion.

A lot of countries in the world have already taken a number of steps to align their financial systems with sustainable development and address

Table 4.3   The green finance policy implementation

Country	Sphere of Implementation	Content
Brazil	BOVESPA Stock Exchange	Has set up a Corporate Sustainability Index as early as 2005.
	Brazil Central Bank	Has introduced requirements for banks to monitor environmental risks, building on a voluntary Green Protocol from the banking sector
	Brazil's banking association	Is developing a standardized assessment methodology and automated data collection system to monitor flows of finance for green economy sectors.
China	People's Bank of China	Introduced green bond standards and green banking regulation
France	French government	Introduced mandatory climate-change-related reporting for institutional investors starting in January 2016.
Germany	German national Development Bank	Emitting Green Bond issuers worldwide.
United Kingdom	Bank of England's Prudential Regulation Authority	Published a report on the impact of climate change on the UK insurance sector.
India	The Reserve Bank of India	Lending to small renewable energy projects.
South Africa	Johannesburg Stock Exchange	Environmental, social and governance disclosure indicators have been introduced

*Compiled by authors on the base of (Bhattacharya et al., 2016).

risks related to climate change (Chygryn, 2017). Also given the diversity of financial systems across the countries, it is clear that measures need to be tailored to the specific needs and circumstances of each country. The compilation of experience on the green finance policy for countries with different financial systems is shown in Table 4.3.

According to the findings Ukraine has already started to develop the relevant mechanism for supporting green investment market development. Thus, at the national level, the Government declared the concept of green bond market development. Green financial instruments with the correct target use, risk assessment could solve the range of issues, in particular: expanding funding for energy projects, strengthening the country's economic potential,

Figure 4.4   The main options for green investment in Ukraine.

*Compiled by the authors on the basis of (Krasnyak et al., 2015; Inderest et al., 2012; Bhattacharya et al., 2016).

and further integrating into the global economic environment. The results of analysis allow allocating the main spheres for green investment in Ukraine (Figure 4.4).

In addition, the G20 experience showed that green investing has a range of opportunities and challenges for Ukrainian investors which justified the developing of supporting mechanism to overcome these barriers and challenges with the purpose to achieve green growth (Figure 4.5).

Ukraine has focused on attracting investments to projects related to renewable energy and energy efficiency both on a national and regional level.

Thus, the Ukrainian Government has to develop priorities and choose strategies of investment in order to stimulate the involvement of green foreign investment and develop a favourable climate for domestic investors. An important aspect of the effective functioning of green investment mechanism is the system of the motivation of the investors' involvement. The attitude of green investment is determined by many motives in their various combinations, which in their turn determine the system of their economic interest. Thus, it is necessary to emphasize and explain the main benefits of green investment for Ukrainian business sector. According to the EU experience, green investment allows receiving not only ecological effect but also social, political and economic. So, the main economic benefits could be as follows:

  – Increase productivity through the use of innovative and environmental technologies and equipment;

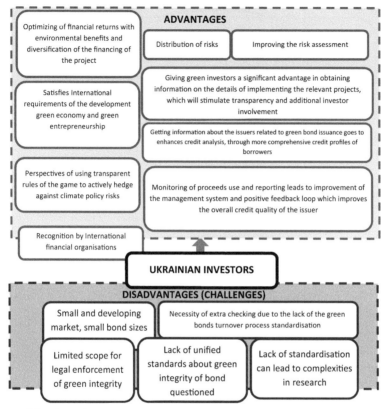

Figure 4.5 Green investment: advantages and disadvantages for investors in Ukraine*.

*Compiled by the authors.

– Reducing costs and product cost based on the reduction of energy intensity and resources;
– Increase the competitiveness business entity and the possibility of entering new markets etc.

From the political point of view: reducing the level of political dependence on foreign suppliers' resources; widening the opportunities to use of international agreements for activation quota trading, green production and etc.

The scientists (Ambec et al., 2008) describe seven channels through which green investments can raise the benefits of firms or cut their costs:

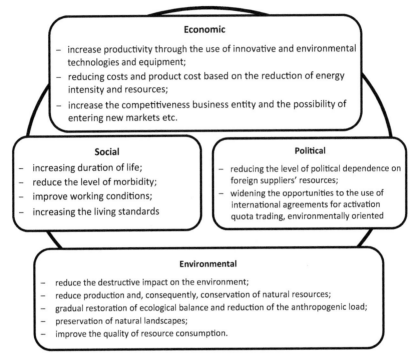

Figure 4.6   The benefits for issuers (who will disburse green investment)*.

*Developed by the authors on the basis of Chygryn, 2015.

better access to markets; possibility for differentiation of products; commercialization of pollution-control technology; savings on regulatory, material, energy and services; capital, and labour costs.

The consolidated benefits for green investors are presented in Figure 4.6.

It is necessary to emphasize policy spheres where the main changes should be done.

1. Promotion the standardization of green finance practices, which includes recognizing the diversity of financial systems, establishing markets for green financial assets, developing principles and guidelines for green finance for all asset classes, including bank credit, bonds and secured assets.

2. Enhance the transparency of information by promoting disclosure standards for carbon and environmental risks: widening disclosure standards for carbon and environmental risks and related information flows for

Table 4.4 Instruments of encouraging green investment*

Level	
National Government	Private
1. Eliminating subsidies for fossil fuels and internalize their social and environmental costs.	1. Providing a one-year grace period for principal repayment, and to reduce transaction fees paid by private developers for project assessment.
2. Establishing long-term binding targets for renewable energy with explicit paths to achieve them to make the opportunity for investment clear.	2. Encouraging local banks to lend to renewable energy projects, provide credit guarantees, risk sharing, and long-term lines of credit.
3. Using international best practices in economic assessment s and offering tenders.	3. Providing technical assistance to develop public underwriting criteria for renewable energy projects.
4. Communicating with investors and other stakeholders to ensure policies and processes address barriers and encourage green investment.	4. Providing accessible and inexpensive country risk insurance to increase the number of private international investors.
	5. Developing structures that encourage new players to finance renewable energy in emerging markets.

*Developed by the authors on the basis of (Chapman, 2016).

addressing the problem of information asymmetry. Because investors often do not know to what extent specific sectors and companies have been affected by climate change.

3. Support market development for green investment: advancement of financial institutions and regulatory frameworks which can play in developing new markets (the market for green bonds).

4. Maintenance developing countries in developing and implementing nationally sustainable finance roadmaps: devising their own national green finance practices and frameworks. Developing sustainable finance roadmaps should be geared to country-specific circumstances and needs and be supported through technical assistance.

The systemised general instruments of encouraging green investment for government and private level are in Table 4.4.

It is also important to note that green investment has a wide sphere of implementation in the different level of the economy. The worldwide experience (Green, 2017; Heinkel et al., 2001; Eyraud et al., 2013) proved the efficiency of developing the green financing market as a way to attract the additional finance recourse to the economy. Thus, for that purpose, the findings showed, that most effective activities are: to organize equity investment funds, to invest in clean cities, to support energy efficiency partnerships with financial institutions, to warehouse energy efficiency loans, bond issuance by development authorities for energy efficiency, to promote innovation at early-stage companies and projects, to overcome financing barriers for residential solar project developers, public lending to facilitate commercial financing for biogas, to finance waste-to-energy, to extend the green bank model to international activities etc. Moreover, the contribution to the growth of the green economy could be realized by cross-industry collaboration, evolving standards and promoting green finance. All green investment initiatives should contribute to a recognized green purpose, reduce of global greenhouse gas emissions, enduring green impact, clear investment criteria implementation, robust green impact evaluation, effective monitoring and engagement, transparent reporting.

4.4 Conclusion

The general increased interest in climate change and environment problem has increased attention to investments in green technologies and sustainable practices. For this reason, the important economic relationships, challenges, perspectives and investment opportunities related to renewables and other green technologies were investigated. The findings allow making the conclusion that green investment is a very wide category and it is being used at all levels: the investment in primary technologies and projects and also to green companies and financial assets. Green investment can be independent, a sub-set of a broader investment theme or closely related to other investment approaches such as socially responsible investing, environmental, social and governance investing sustainable, long-term investing and others. The investors' and financial institutions' attention to climate change and environmental problems in general, has been rising in recent years and investor financial initiatives in this respect are growing also. It is important to note that energy efficiency and environment protection represent a significant largely untapped opportunity for meeting the dual goals of risk-adjusted financial return and environmental protection. The world financial institutions are interested to invest in renewables as a cost-effective opportunity to reduce the carbon emissions and to prevent environmental damage. Thus, essential to develop specific recommendations for government and private investors for "green investing", which should include: encouraging consideration of grcen standards for all levels of the investment decision-process; transparency in green issues? and strengthening disclosure to consumers and investors; encourage capacity building and development of internal and external 'green audit' as well as raising 'green' awareness and education. Ukraine will need to prioritize activities in investing in renewables and green technologies by selecting its own criteria, perhaps in conjunction with potential investors of a project-by-project basis.

Acknowledgment

This research was funded by the grant from the Ministry of Education and Science of Ukraine (№ g/r 0117U003932).

References

Adeel-Farooq, R. M., Abu Bakar, N. A. and Olajide Raji, J. (2018). Greenfield investment and environmental performance: A case of selected nine developing countries of Asia. Environmental Progress & Sustainable Energy, 37(3), 1085–1092.

Ambec, S. and Lanoie, P. (2008). Does it pay to be green? A systematic overview. Academy of Management Perspectives 22(4), 45–62.

Andreeva, N. M. (2005). The theoretical basics of ecologization of investment activity. Research Bulletin of the National Forestry University, 15, 314–320.

Anischenko, V. O. (2007). To the issue of improving the theoretical and methodological principles of environmental investment. Actual problems of economics, 8, 175–183.

Arestov, S. V. (2010). Foundations of ecosystem transfer in environmental investment. Retrieved from: http://www.nbuv.gov.ua/portal/soc_gum/en_re/2010_7_2/2.pdf.

Bhattacharya, A., Meltzer, J. P., Oppenheim, J., Qureshi, Z. and Stern, N. (2016). Delivering on sustainable infrastructure for better development and better climate. Brookings Institution. Available: https://www.brookings.edu/wp-content/uploads/2016/12/global_122316_delivering-on-sustainableinfrastructure.pdf.

Bloomberg New Energy Finance. Retrieved from: https://about.bnef.com/clean-energy-investment/.

Cebula, J. and Pimonenko, T. (2015). Comparison Financing Conditions of The Development Biogas Sector in Poland and Ukraine. International Journal of Ecology & Development™ 30(2), 20–30.

Chapman, Sarah., Inouye, Lauren., Smith, James. and Toriello, Carlos. (2016). Infrastructure for sustainable development: Central America renewable energy case study. IWANA.

Chigrin, O. and Pimonenko, T. (2014). The ways of corporate sector firms financing for sustainability of performance. International Journal of Ecology & Development™ 29(3), 1–13.

Chygryn, O. Y. (2015). Ways to financing environmental and recourse saving activity in Ukraine. Sustainable Human Development of local community. Civil Society, 278–284.

Chygryn, O. (2016). The mechanism of the resource-saving activity at joint stock companies: The theory and implementation features. International Journal of Ecology and Development, 31(3), 42–59.

Chygryn, O. (2017). Green entrepreneurship: EU experience and Ukraine perspectives. Waste management, 243(5), 146.

Chygryn, O., Petrushenko, Y., Vysochyna, A. and Vorontsova, A. (2018). Assesment of fiscal decentralization influence on social and economic development. Montenegrin Journal of Economics, 14 (4), 69–84.

Chygryn, O., Pimonenko, T., Lyulyov, O. and Goncharova, A. (2018). Green Bonds Like the Incentive Instrument for Cleaner Production at the Government and Corporate Levels: Experience from EU to Ukraine. Journal of Environmental Management and Tourism, (Volume 8, Winter), 9(17): 105–113. doi: 10.14505/jemt.v5.2(10).01.

Climate Transparency (2017). Brown to Green: The G20 Transition to a Low-Carbon Economy, Climate Transparency, c/o Humboldt-Viadrina Governance Platform, Berlin, Germany, www.climate-transparency.org.

Djalilov, K., Vasylieva, T., Lyeonov, S. and Lasukova, A. (2015). Corporate social responsibility and bank performance in transition countries. Corporate Ownership and Control, 13(1CONT8), 879–888.

Eurosif, (2011). Corporate Pension Funds & Sustainable Investment Study.

Eyraud, L., Clements, B. and Wane, A. (2013). Green investment: Trends and determinants. Energy Policy, 60, 852–865. doi: 10.1016/j.enpol.2013.04.039.

Green Investment Banks Innovative Public Financial Institutions Scaling up Private, Low-carbon Investment (2017). OECD Environment Policy Paper No. 6, January. Retrieved from: Green Investment. Retrieved from: http://wgeco.org/green-investment/.

Hagspiel, V., Dalby, P. A. O., Gillerhaugen, G. R., Leth-Olsen, T. and Thijssen, J. J. J. (2018). Green investment under policy uncertainty and Bayesian learning, Energy, doi: 10.1016/j.energy.2018.07.137.

Heinkel, R., Kraus, A. and Zechner, J. (2001). The Effect of Green Investment on Corporate Behavior. *Journal of Financial and Quantitative Analysis*, 36(4), 431–449. doi: 10.2307/2676219. https://www.oecd-ilibrary.org/docserver/e3c2526c-en.pdf?expires=155 2237619&id=id&accname=guest&checksum=39502C79D3DA4B93EF 51D97E0771A868.

Ibragimov, Z., Lyeonov, S. and Pimonenko, T. (2019). Green investing for SDGS: EU experience for developing countries. Economic and Social Development (Book of Proceedings), 37th International Scientific Conference on Economic and Social Development – "Socio Economic Problems of Sustainable Development", Baku, pp. 867–876.

Inderst, G., Kaminker, C. and Stewart, F. (2012). "Defining and Measuring Green Investments: Implications for Institutional Investors' Asset Allocations", OECD Working Papers on Finance, Insurance and Private Pensions, No. 24, OECD Publishing. http://dx.doi.org/10.1787/5k9312t wnn44-en.

Inderst, G. (2016). Infrastructure Investment, Private Finance and Institutional Investors: Asia from a Global Perspective. ADBI Working Paper Series, No. 555, January. Manila: Asian Development Bank Institute.

International Energy Agency (2016). International Energy Outlook 2016. https://www.eia. gov/outlooks/ieo/pdf/0484(2016).pdf.

Investment. Retrieved from: https://www.investopedia.com/terms/i/investm ent.asp#ixzz5LVfUJt3b.

Investopedia (2018). Green Investing. Retrieved from: https://www.investop edia.com/terms/g/green-investing.asp#ixzz5La3ZIXli.

Krasnyak, V. and Chygryn, O. (2015). Theoretical and applied aspects of the development of environmental investment in Ukraine. *Marketing and Management of Innovations*. 3, 226–234.

Kvaktun O. O. (2014) The real green investments as an effective tool for sustainable design and construction of Ukraine's regions. Retrieved from: http://ecoukraine.org/_ld/0/7_ecpros_2014_83.pdf.

Lyeonov, S. V., Vasylieva, T. A. and Lyulyov, O. V. (2018). Macroeconomic stability evaluation in countries of lower-middle income economies. *Naukovyi Visnyk Natsionalnoho Hirnychoho Universytetu*, 1, 138–146. doi: 10.29202/nvngu/2018-1/4.

Martin, P. R. and Moser, D. V. (2016). Managers' green investment disclosures and investors' reaction. Journal of Accounting and Economics, 61(1), 239–254. doi: 10.1016/j.jacceco.2015.08.004.

Mielke, J. and Steudle, G. A. (2018). Green Investment and Coordination Failure: An Investors' Perspective. Ecological Economics, 150, 88–95. doi: 10.1016/j.ecolecon.2018.03.018.

Novethic (2011). Survey, European Asset Owners? ESG Perceptions and Integration Practices.

Novethic (2012). Green Funds. A sluggish Market.

OECD (2006). Infrastructure to 2030: Telecom, Land transport, Water and Electricity. Paris.

OECD (2008). Is it ODA? Paris: OECD. Available at: http://www.oecd.org /dac/stats/34086975.pdf.

OECD (2012). Strategic Transport Infrastructure Needs to 2030. International Futures Programme. Paris: OECD.

OECD (2013). Long-Term Investors and Green Infrastructure: Policy Highlights. Paris: OECD. http://www.oecd.org/env/cc/Investors%20in%20G reen%20Infrastructure%20brochure%20(f)%20[lr].pdf.

OECD (2013). What do we know about Multilateral Aid: The 54 billion dollar question. OECD Policy Brief. Paris. http://www.oecd.org/development/financing-sustainable-development/ 13_03_18%20Policy%20Briefing%20on%20Multilateral %20Aid.pdf.

OECD (2014). OECD Science, Technology and Industry Outlook. Paris: OECD.

OECD (2014). Pooling of Institutional Investors Capital – Selected Case Studies in Unlisted Equity Infrastructure. Paris: OECD.

OECD (2014). Private Financing and Government Support to Promote Long-Term Investments in Infrastructure, p. 36. Paris: OECD.

OECD (2014). Report on Effective Approaches to Support Implementation of the G20/OECD High-Level Principles on Long-Term Investment Financing by Institutional Investors, p. 30. Paris: OECD.

OECD (2014a). Infrastructure to 2030: Mapping Policy for Electricity, Water and Transport. Paris: OECD.

OECD (2015). Climate Finance in 2013–15 and the USD 100 billion goal. Report by OECD in collaboration with CPI, p. 10. Paris: OECD.

OECD (2015a). Towards a Framework for the Governance of Public Infrastructure. OECD Report to G20 Finance Ministers and Central Bank Governors. Paris: OECD.

OECD (2015b). G20/OECD Report on Investment Strategies. Rreport prepared for G20 Finance Ministers and Central Bank Governors. Paris: OECD.

OECD (2015c). Size of Public Procurement. In Government at a Glance. Paris: OECD.

OECD (2015d). Smart Procurement: Going Green – Better Practices for Green Procurement. GOV/PGC/ETH(2014)1/REV1. Paris: OECD.

OECD (2015e). Taxing Energy Use: OECD and Selected Partner Countries. Paris: OECD.

OECD (2015f). Mapping of Instruments and Incentives for Infrastructure Financing: A Taxonomy. OECD Report to G20 Finance Ministers and Central Bank Governors, September 2015. Paris: OECD.

OECD (2015g). Effective Approaches to Support Implementation of the G20/OECD High-Level Principles on Long-Term Investment Financing by Institutional Investors. Report to G20. Paris: OECD.

OECD (2015h). Mapping Channels to Mobilize Institutional Investment in Sustainable Energy. Paris: OECD.

OECD (2016). DAC Members List. Available at: http://www.oecd.org/dac/dacmembers.htm.

Pimonenko, T., Prokopenko, O. and Dado, J. (2017). Net zero house: EU experience in ukrainian conditions. *International Journal of Ecological Economics and Statistics*, 38(4), 46–57.

Prokopenko, O., Cebula, J., Chayen, S. and Pimonenko, T. (2017). Wind energy in Israel, Poland and Ukraine: Features and opportunities. *International Journal of Ecology and Development*, 32(1), 98–107.

Summary of Stream 5: Our green future: green investment and growing our natural assets. Retrieved from: http://www.fao.org/fileadmin/user_upload/rap/Asia-Pacific_Forestry_Week/doc/Stream_5/Stream_5_Summary.pdf.

The 2030 Agenda for Sustainable Development. Retrieved from: https://sustainabledevelopment.un.org/content/documents/21252030%20Agenda%20for%20Sustainable%20Development%20web.pdf.

The Green Investment Report (2013). World Economic Forum. www3.weforum.org/docs/WEF_GreenInvestmentReport_ExecutiveSummary_2013.pdf.

Triodos Bank. Green investment – what does it actually mean? Retrieved from: https://www.triodos.co.uk/en/personal/ethical-investments/green-investments/.

Unlocking the green bond potential in India. Retrieved from: https://archiv
e.nyu.edu/bitstream/2451/42243/2/Unlocking%20the%20Green%20B
ond%20Potential%20in%20India.pdf. Green bonds as a bridge to the
SDGs (2018). Retrieved from: https://www.climatebonds.net/files/files
/CBI%20Briefing%20Green%20Bonds%20Bridge%20to%20SDGs%2
81%29.pdf. The green bond market in Europe. (2018). Prepared by the
Climate Bonds Initiative. Retrieved from: https://www.climatebonds.
net/files/files/The%20Green%20Bond%20Market%20in%20Europe.
pdf.

Vasylyeva, T. A. and Pryymenko, S. A. (2014). Environmental economic
assessment of energy resources in the context of ukraine's energy
security. Actual Problems of Economics, 160(1), 252–260.

Vyshnitskaya, O. I. (2013). Environmental investments: essence, classification, principles and directions of realization – Bulletin of Sumy State
University. Economy Ser., 2, 51–58 320.

Yevdokimov, Y., Chygryn, O., Pimonenko, T. and Lyulyov, O. (2018).
Biogas as an alternative energy resource for ukrainian companies: EU
experience. Innovative Marketing, 14(2), 7–15. doi: 10.21511/im.14(2).
2018.01.

5

Reaching SDGs Through Public Institutions and Good Governance: Why Trust Matters

Anna Buriak* and Oleksandr Artemenko

Department of Finance, Banking and Insurance,
Sumy State University, Ukraine
E-mail: ann.v.buriak@gmail.com
*Corresponding Author

This chapter discusses the importance of trust as a key determinant of good governance and potential for growth and sustainable development in the context of reaching SDGs. It has been highlighted that implementation of SDGs requires both trust in public institutions – for better compliance with regulatory initiatives and trust in other people – for more cooperative behavior and collaborative actions towards sustainable solutions. This chapter encompasses conceptual framework for understanding trust as emotional determinant of economic and social processes of the country. Since huge decline of trust in public institutions has been fixed during global financial crisis of 2008, author focuses on multilevel concept for restoring trust in financial system of the country. As building effective institutions in the context of SDGs implementation requires development of accountable and transparent institutions at all levels so case of ensuring integrity in financial system of Ukraine is presented by this chapter.

5.1 Introduction

The success of Sustainable Development Goals (SDGs) crucially depends on the good public governance and efforts from public institutions. SDG 16 promotes "access to justice for all and build effective, accountable and inclusive institutions at all levels" highlighting crucial role of public institutions in reaching SDGs. In turn it requires public trust in institutions as well as interpersonal trust among citizens. Trust is important for implementation of many government policies and programmes influencing on behavioural responses from the public.

Recent OECD's report "Trust and Public Policy. How Better Governance Can Help Rebuild Public Trust" examines the influence of trust on policy making through two components of trust – competence and values with regarding its relevant dimensions like responsiveness, reliability, integrity, openness, fairness – to strengthen public trust and effective implement of reform agenda (OECD, 2017).

According to the "Compendium of innovative practices in Public Governance and Administration for Sustainable Development" presented by United Nations the intension of sustainable development will require more collaborative actions of stakeholders towards sustainable solutions through collaboration across public, private and civil society sectors (UN, 2016). Bringing together stakeholders to get sustainable outcomes is undoubtedly associated with cooperation and compliance of citizens and depend on interpersonal trust in the country (usually measured by generalized trust) as key determinant a country's potential for growth and sustainable development.

Trust is an important issue for both academic research and policy-making. For example, trust is regarded as measure for monitoring human well-being in the context of OECD Better Life Initiative (OECD, 2013). It is highlighted that trust in other people and trust in institutions are essential elements for economic and social progress of the country including implementation of SDGs.

In this chapter, we show that huge progress has been made on trust issue in mainstream economics as well as other social disciplines. Firstly, we present conceptual framework for understanding trust as emotional determinant of economic and social processes of the country and across the world. Besides theoretical backgrounds address multifaceted nature of trust, main dimensions and role in the society's progress. Secondly, multilevel concept for restoring trust in financial system of the country is outlined since global financial crisis has been associated with trust crisis and the loss of financial and other public institutions' reputation, collapse of public confidence and trust. At the end, case of ensuring integrity in financial system of Ukraine is presented by this chapter in the context of building accountable, transparent and effective institutions at all levels to implement SDGs.

5.2 Trust as Emotional Determinant of Economic and Social Progress of the Country: Foundations and Overview

The global financial crisis of 2007–2008 and following long-term recession has been contributed to a significant revision of both economic and social country development' determinants. The high level of uncertainty and low level of confidence in the future has been changed crucially the behavior of economic agents weakened the role of fundamental determinants and strengthened the influence of psychological (emotional) origin.

The emotional determinant has become widespread with the term "sentiments" defined as "attitude, opinion or judgment based on feelings" and is used to describe the views of the economic agent on future economic and social development. In turn it is reflected in the current economic decisions and influences future economic dynamics. Typically, this determinant is reflected through the waves of optimism and pessimism and is determined by surveys of households and firms about their personal opinion about future economic and social development.

Main fields of sentiments' impact are associated with consumption and volatility of business cycles including the processes of investment and production. According to theoretical and empirical evidence significant shocks of uncertainty have a considerable impact on the real sector due to the substantial changes in the behavior of economic agents provoking decline of state regulation policy effectiveness (ECB, 2013). To explain transmission of the emotional determinant to economic behavior there are two approaches used in the literature – rational which defines the emotional determinant as a component of the imperfect information field for making economic decisions and – irrational associated with irrational instincts (animal spirits) (ECB, 2017). At the same time, the concept of "confidence" as a kind of emotional determinant is used from the standpoint of a rational (informational) approach and based on objective information, while the concept of "trust" is predominantly based on the subjective information due to the uncertainty of the economic and social environment and, consequently, irrational behavior (Tonkiss, 2009; Akerlof and Shiller, 2009).

Principles of psychological science explain content of "trust" by the presence of two components – cognitive (contains objective knowledge about the world and norms of behavior) and emotional (contains subjective characteristics, individual feelings, emotions). In sociology trust is considered as a characteristic of human activity in a "society of risk" as a component of social capital (Fukuyama, 1995).

The global financial crisis has been associated with trust crisis featuring credit freeze at many financial markets, the loss of financial institutions' reputation, lack of transparency in financial reporting, collapse of public confidence and trust (Schatz and Watson, 2011; Roth, 2009; Gros and Roth, 2010; Sapienza and Zingales, 2012). Trust becomes crucial for economic dynamics and financial market activity when legal enforceable contracts are absent and confidence in market structures is undermined.

Trust is considered to be multidimensional, heterogeneous and multilevel category. Overview of the recently made progress in economic, social and psychological research of trust topic give us foundations to assert the existence of three components of trust – interpersonal, institutional and systemic linked by close relationships between the indicated levels (components) and existence of interactions between them.

5.2.1 Interpersonal Trust

Interpersonal component of trust reveals psychological reasons of agent's behavior (Simpson, 2007; Borum, 2010). Trust as an individual's belief or the expectancy held by an individual or group (trustor) influences interactions among individuals, groups and organizations.

The level of interpersonal trust in society is also of interest in the context of economic, political and social research as productivity's determinant and a factor affecting a variety of fields including health, crime, life satisfaction etc.

From economic point of view interpersonal trust is considered as an integral part of social capital in society (Fukuyama, 1995; Putnam, 1993) which defines performance of public institutions (La Porta et al., 1997), economic growth and financial development (Guiso, Sapienza and Zingales, 2008). The impact of social capital on financial development is enhancing through financial contracts as trust-intensive contracts. Findings for Italy region show that social capital plays an important role in the financial development and seems to matter the most when education levels are low and law enforcement is weak.

Zak and Knack (2001) reveal that interpersonal trust substantially impacts economic growth by reducing transactions costs among economic actors driven by asymmetric and costly information. Moreover, these findings form ground for trust-raising policies to stimulate economic growth (Zak and Knack, 2002). Applying interpersonal trust to the financial system indicators it is empirically proved that one fifth of the variation in intra-euro area economic imbalances (a combination of the government budget balance, the inflation rate and the current account balance) are due to differences in interpersonal trust (Bützer, Jordan and Stracca, 2013).

Formal institutions that enforce contracts (rule of law, bureaucracy, etc.), social and economic heterogeneity, wealth and income are served as determinants of interpersonal trust among citizens. Interpersonal trust in the country is formed under the influence of factors of the institutional, social and economic environment. Investigation of trust relation to macro-economic performance for 33 countries (Ruben, 2013) introduced negative effect that of the modernization on interpersonal relations and positive one of a society's education, as well as through diminishing linguistic barriers. Despite the significant influence of cultural, historical, ethnic, religious, ideological norms on the level of interpersonal trust, the most important one is economic conditions such as the level of wealth of the country and the income distribution in society (Knack and Keefer, 1997).

There are two conceptual approaches to measuring interpersonal trust in society – experimental/laboratory (trust games) and sociological surveys. In sociological studies, the traditional indicator of interpersonal trust is Rosenberg's question (1957) – "Do you believe that you can trust most people?" – used in the World Values Survey (WVS). Authors of the research proved the equivalence of these approaches in cross-border research and the expediency of using the indicator of generalized trust as an indicator of interpersonal trust in society (Carlin et al., 2017). A global review of values also gives an opportunity to evaluate specific trust (particularized trust) – in relation to certain social groups: the family, neighbors, acquaintances and strangers, people of another religion and nationality. If interpersonal trust is trust at a narrow level, system trust is at the broad one.

5.2.2 System Trust

Another crucial contribution of trust to economic and social development is made by systemic level reflecting trust to economic system in general and its main elements – banking sector, payment system, stock market, interbank market and infrastructure like accounting and auditing standards, a system of regulation and supervision, credit and rating agencies etc. (Sapienza and Zingales, 2012). In terms of the financial system the key indicator of the trust level is price and financial stability. This point confirms that reform of the control mechanisms in the financial sector is one of the most common areas for restoring system trust such as the creation of a single supervisory mechanism in the European banking system (Fleck and von Lüde, 2015).

5.2.3 Institutional Trust

One of the reasons of emergence of the institutional trust concept is the ineffectiveness of the mechanism of interpersonal trust under information asymmetry and the need to create specialized "expert" institutions to minimize the role of personal relationships and the need for the existence of interpersonal trust. In this case financial intermediaries are institutions of delegated monitoring. Institutional trust is formed on the basis of meeting the expectations of the consumers of services and products of a particular institute with the results/processes of such consumption in particular in the financial market. In the economic literature there are several components (determinants) that build trust in the institution, with the composition and influence of these determinants of institutional trust varied in various studies. Therefore, link between trust and economic and social development of the country is obvious and empirically proved but at the same time is not straightforward. One of explanations could be found in period of business cycle (stages) – during non-crisis periods there is dominant role of the rational side of trust – related to fundamental factors and during crisis periods with collapse in confidence prevailed role is given to irrational side of trust – animal spirits.This viewpoint underlines pro-cyclical character of trust indicators based on fundamental as well as irrational forces.

Good governance and effective implementation of SDGs require institutional trust – trust in public institutions (including government). Since financial and economic crisis of 2008 led to a significant loss of trust in government ("... by 2012, on average only four out of ten people in OECD member countries expressed confidence in their government ..." (OECD, 2013), issue of trust in public institutions has become of high importance. Trust in governments, markets and other institutions is essential ingredient for public sector reform agenda provoking high compliance with regulations, defining legitimacy and sustainability of political systems, enhancing human well-being and policy effectiveness. According to the Edelman Trust Barometer, which measures trust on a scale of 0–100, three-quarters of governments around the world are distrusted by their citizens and "... the last decade has seen a loss of faith in traditional authority figures and institutions ...". Government (48 percent) is less trusted than NGOs (57 percent), business (56 percent) and employer (75 percent) (Edelman Trust Barometer,

2018). Moreover, relying on Gallup's survey about public confidence in various institutions, declining trust is considered to be from 1995.

Financial services, despite slight increase to 57 percent, was once again the least-trusted globally – for example, trust in technology sector in 2018 was 78 percent, manufacturing – 70 percent and automotive – 70 percent. Taking it in consideration and since huge decline of trust in public institutions has been fixed during global financial crisis of 2008, further author focuses on multilevel concept for restoring trust in financial system of the country.

5.3 Multilevel Concept for Restoring Trust in Financial System of the Country

The financial system is regarded as a complex socio-economic system that involves the interaction of both economic agents (individuals) and institutions (including social, political, legal, ethical). This indicates the importance of combining several methodological approaches for the research: (1) behavioral – at the level of economic agents, which allows them to take into account their individual psychological and emotional drivers in decision making; (2) institutional – at the institutions' level which takes into account the peculiarities of the functioning and interaction of social and economic institutions and systems. The multilevelness of trust in financial system is becoming of great importance during times of it restoring. Chairman of the Financial Stability Board emphasizes the need to differentiate between the following levels of trust – between banks and their shareholders, borrowers, supervisors; between supervisors in developed and developing countries (Carney, 2013).

Multilevel nature of trust is basis of the global financial crisis study by Gillespie and Hurley (2013), according to this crisis is followed by loss of confidence in main financial system agents – individual agents (investors, employees), organizational institutions (Lehman Brothers, Citibank), individual markets (financial, insurance ones) and individual representatives of public interests (policymakers, governments). Multilevel concept for restoring trust in financial system of the country is based on sources of public trust differentiating three levels of financial system research – interpersonal, institutional and system (Figure 5.1). The first source (level) of public trust in the financial system is the interpersonal trust between economic agents (known as generalized trust) and characterizes their willingness to establish agreements and cooperation which in turn reduces transaction costs and increases economic efficiency of policy-making. From the economic standpoint the concept of social capital is of high importance allowing to investigate the trust

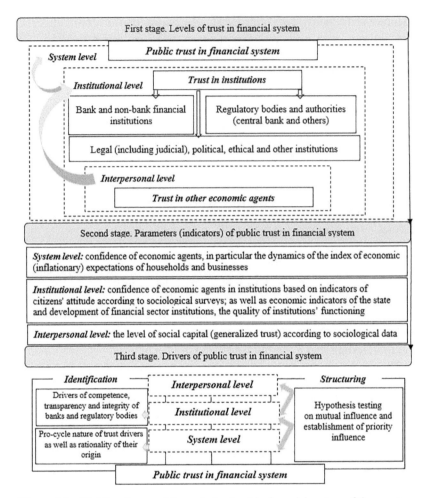

Figure 5.1   Multilevel concept for restoring trust in financial system of the country.

between economic individuals as a public resource that has a positive impact on the economic indicators of the country's development. The application of the concept of social capital in the research of public trust allows to take into account economic motives of the individuals as well as socially motivated motivation of economic agents based on habits and traditions.

The next source (level) of research of public trust in the financial system is the institutional characterizing the expectations of economic agents and their attitude to the institutions – in particular, the financial market, public authorities, and others. Trust in such institutions is formed on the basis of confidence in the stable performance of the functions assigned to them. Consequently, the stable and qualitative level of the functioning of such institutions creates the confidence of economic agents, and as a consequence, the trust that directly determines their future behavior.

The third source (level) of research of public trust in the financial system is system, resuled from the complex interaction of institutional and interpersonal levels of trust and is characterized by confidence in the functioning of the financial system. In the context of the financial system, price and financial stability as key targets of modern central banks and financial authorities, are of the high importance.

Identification pf parameters (indicators) of public trust in financial system is obligotary to identify and structure the drivers of restoring public trust in the financial system. At the system level trust indicator is considered to be the confidence of economic agents, in particular the dynamics of the index of economic (in particular, inflationary) expectations of households and businesses; at the institutional level – trust of economic agents in institutions based on indicators of citizens' attitude according to sociological surveys as well as economic indicators of the state and development of institutions of the financial system, the quality of institutions' functioning; on interpersonal level – the level of generalized trust according to sociological data.

Financial markets' development and creation of new financial institutions, instruments and technologies has provoked substantial shift in relations between economic agents and institutions – the role and significance of interpersonal trust is minimizing, and the role of institutions and structures of the financial system is increasing including the importance of regulatory systems in ensuring the financial stability. Identification of interpersonal trust's drivers (the first level and source of restoration of public trust in the financial system) has psychological context and is the field of research by psychology and sociology. Drivers of the institutional and system levels of research of public trust in the financial system are of the more importance for identification.

Institutional trust is determined by different factors like competence (reflects the level of its expertise in the marketing, financial, managerial field, etc.), transparency (openness for all stakeholders, including information not only about activities, but also products and services), integrity (involves ethical relations with stakeholders, a community of values and attitudes, and reflected by the level of social responsibility to society). According to the results of structural equations modeling in the Netherlands integrity is the most important determinant of bank trust as well as transparency, customer orientation, and competence are significant also (Pauline, 2017). At the system level the identification of public trust's drivers involves identification the cyclicality of trust and its main determinants as well as the rational/irrational nature of their origin by testing the hypotheses of rational/irrational expectations theory.

Structuring the drivers of public trust in the financial system involves the research of mutual influence and relationships between levels and their relevant drivers. Results of the research project on trust and confidence in the financial sector of the Netherlands have shown a positive relationship between interpersonal trust and confidence in Parliament, the central bank, business in general, banks and non-bank financial institutions, the national unit of the euro (Mosch and Prast, 2008). At the same time, the direction of interconnection remained uncertain: (1) high level of trust in society stimulates transparent and open behavior of banks and thus forms a high level of trust in them or (2) high level of trust in the institutions of the country forms a trusting behavior among other economic agents. Therefore, empirical verification requires the identification of the relationship: first, trust in financial institutions, in particular banks, and interpersonal trust among citizens; secondly, confidence in the financial institutions, in particular banks, and the trustworthiness of the state as a whole and courts as institutions protecting the rights of creditors and consumers of financial services.

5.4 Trust and Integrity – Case of Financial System of Ukraine

Building effective institutions in the context of SDGs implementation requires development of accountable and transparent institutions at all levels so case of ensuring integrity in financial system of Ukraine is presented further.

Ukrainian financial system reform originated in 2014 in the wake of the signature of the EU-Ukraine Association Agreement was regarded as one of the cornerstones of the efforts to rebuild the national economy along European lines (National Bank of Ukraine, 2016). At the start of reform main issues of the Ukrainian financial system included: high level of dollarization of credits and deposits, severe administrative foreign exchange restrictions, the mass exodus of large European financial-sector players from the Ukrainian market (about 10 European banks sold their banking subsidiaries in Ukraine or stopped doing retail business), and a drastic decline in both nominal (-20%) and real (-40%) lending growth (Buriak, 2017).

Related-party loans were more prevalent in small- and medium-sized regional banks with assets of less than $10 billion, which tend to have established shareholder-linked client bases and operate in geographically restricted areas of low business diversification. So, the basic principles of financial reform include balancing economic interests with the integrity of the financial system through the comprehensive protection of creditor, consumer and investor rights. One of the goals of the reform process in Ukraine related to raising the standards of information disclosure to the benefit of consumers and financial sector investors.

Therefore, one of the institutional deformations of the Ukrainian financial system is insufficient level of transparency growing with the loss of public confidence due to the numerous bankruptcies of bank institutions, the insecurity of depositors' rights, the low level of transparency of domestic bank institutions regarding the ownership structure as well as implementation banking activities in the interests of bank owners.

According to international methodology – consolidated transparency index of Nayer – level of transparency of the 10 largest banks of Ukraine in recent years can be defined as average (Figure 5.2). Banks do not disclose sufficient information on off-balance sheet items, problem loans, regulatory capital components, non-interest income, and an inadequate classification of loans to counterparties and related parties.

Reduce in the level of transparency in 2014 and 2015 is caused by the lack of disclosure of full or partial information on trading and investment securities, off-balance sheet items and the lack of detailed information on some of the regulatory capital components in some banks. It should be noted that the lowest level of transparency was observed among the banks which were declared insolvent later.

Based on data from the rating agency IBI-Rating it should be noted that small banks during the period of banking sector reform have increased their level of transparency compared with the past years. However, biggest

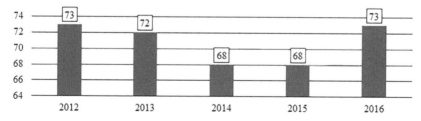

Figure 5.2 Dynamics of the transparency index of the 10 largest banks of Ukraine, %.

bank institutions showed the highest level of progress. The most complete information on quarterly and annual reports, obligatory economic standards of the central bank is provided. This is mainly due to the fact that the National bank of Ukraine imposes mandatory requirements for disclosure of this information.

There is also a tendency for banks to report in accordance with International Financial Reporting Standards. Strengthening of the NBU's requirements for disclosure of information on the ownership structure has increased the level of disclosure in this section of the indicators (up to 81%). In addition, the indicator of disclosure of corporate governance information has improved considerably, in particular, the publication on official sites of the statute and internal documents regulating the work of the governing bodies of the bank as well as providing banks with more detailed information on the authorities and officials. Large banks are more willing to disclose additional information regarding their activities, besides compulsory. Banks with foreign capital in their majority are sufficiently full and timely to disclose information about the state of corporate governance. This approach is explained by long-standing European traditions, but also clearly defined requirements of international standards of transparency. To sum up, it should be noted that the level of transparency of banks in recent years has been improved, but the least developed aspects of the banks are the quality of the published information – availability in terms of the logic of placing on the official websites of banks (only a small amount banks place on the sites consolidated and aggregated data in the form of infographics, presentations, charts, etc.), information on the remuneration policy of the top management of the bank institutions, the results of stress and back testing, as well as full and high-quality information on banking products. A significant contribution to increasing the issue of mistrust of banks is the unreliability or lack of understanding by the consumer of banking services information on the conditions of use of products and services of the bank, especially in terms of risks and prices.

Transparency and integrity in banking is an important aspect of updated missions of financial institutions for sustainability. Ensuring adequate information disclosure for bank clients about main activities and products in an appropriate way will be considered as sustainable practice for modern banking. Since disclosure of information on financial products is lacking sufficient attention during evaluation of bank transparency it is necessary to take into account the component – the transparency of financial products and services of the bank. Proposed approach uses the integral index of bank transparency based on the point rating as percentage of the points sum for all the criteria received by the bank and the maximum number of points that can be obtained bank in accordance with this methodology. Each indicator of disclosure quality is estimated at 1 point, if the information is fully disclosed, sufficiently understandable for the perceptions by clients and relevant at the moment of analysis; 0.5 points – the information is disclosed partly or in an incomprehensible form for the understanding; 0 points is given if there is no information regarding this indicator. Four criteria for bank transparency evaluation were developed:

1. The disclosure of general information about the bank will provide market participants with an idea of the bank institution's mission, the main strategic objectives as well as the place of the bank in the financial services market.
2. Relevant information disclosure of bank's operational activities and financial reporting will allow to assess the quality and effectiveness of the bank's activities, determine both level of bank risk and appetite for risk.
3. Information about ownership structure and corporate governance will allow market participants to familiarize themselves with the principles of doing business by bank owners, the level of stakeholders' influence on the bank performance.
4. Disclosure of reliable and complete information about bank products and services will allow market participants to identify the most optimal product or service appropriate for their needs, and the availability of aggregated reports on product and service conditions, financial calculator and auxiliary videos will help accelerate decision making.

This approach was applied to the 10 largest banks of Ukraine (Table 5.1). The most transparent was the general information about banks and ownership structure. However, banks reveal insufficient information about corporate social responsibility and ratings of bank institutions.

Analysis of bank transparency about financial products and services showed lack of information about full value of products as well as inavailability of auxiliary videos and consolidated reports on the basic conditions of bank products and services. It makes difficult for consumers to understand the total amount of costs for a particular product and does not give the opportunity for quickly and qualitatively comparison of the proposed bank products. Only two of the top 10 bank institutions in Ukraine have certain reports on the list of products and their conditions – it indicates poor quality disclosure by the largest banks in Ukraine. Moreover, banks do not disclose

Table 5.1   Transparency of 10 largest Ukrainian banks for 2012–2017, %*

| Year | Main Criteria of Transparency | | | |
	General Information About the Bank	Disclosure of Bank's Operational Activities and Financial Reporting	Information About Ownership Structure and Corporate Governance	Information About Bank Products and Services
2012	80.7	67.9	77.1	64.2
2013	80.6	68.0	78.3	66.6
2014	81.2	70.0	82.5	67.5
2015	78.7	70.7	85.4	67.5
2016	77.8	69.7	85.8	69.2
2017	75.5	68.3	84.4	70.1

*Own calculations.

all information on costs, such as fees and additional services. Product price information focuses on interest rates and major commissions leading to concealing the value of the product and services by banks. However, it should be noted that banks publish on their websites the terms and conditions for banking services provided by the bank enhancing bank transparency and enabling consumers to familiarize themselves with the basic rights, rules and obligations, contract samples, etc. Therefore, there is a need to increase the transparency of financial products and services of banks through the introduction of new standards and recommendations by the regulator.

Transparency of bank products relating to the disclosure of pricing information and additional terms is the basis for building trust among customers. In turn trust is the foundation for establishing long-term relationships with a bank institution. Bank customers require more transparency about products and services as well as clear explanations of what they are paying for and why.

This issue has attracted recently a lot of attention from international organizations (Basel Committee on Banking Supervision, 2017) academics and business. Thus the World Bank highlights information transparency of financial products and services in the banking sector as one of the direction of consumer protection (The World Bank Group, 2017). Central banks across the world approve regulatory documents for the transparency of financial products and services provided by commercial banks. The purpose of these documents is to define requirements concerning the ways and means of providing information on bank products and services to the client for ensuring transparency and protection of bank institutions' clients. Paper by Totolo et al. (2017), focuses on issues of information disclosure about the full price of bank product or service: (1) the existence of confusion about the actual cost

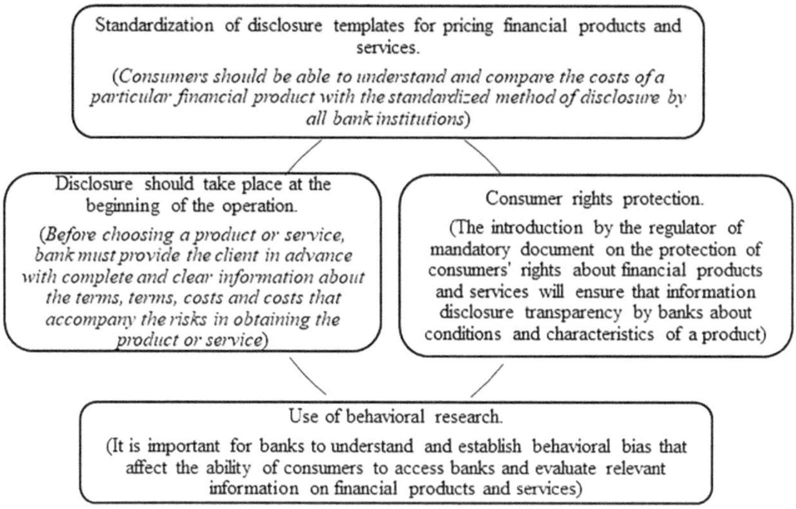

Standardization of disclosure templates for pricing financial products and services.

(Consumers should be able to understand and compare the costs of a particular financial product with the standardized method of disclosure by all bank institutions)

Disclosure should take place at the beginning of the operation.

(Before choosing a product or service, bank must provide the client in advance with complete and clear information about the terms, terms, costs and costs that accompany the risks in obtaining the product or service)

Consumer rights protection.

(The introduction by the regulator of mandatory document on the protection of consumers' rights about financial products and services will ensure that information disclosure transparency by banks about conditions and characteristics of a product)

Use of behavioral research.

(It is important for banks to understand and establish behavioral bias that affect the ability of consumers to access banks and evaluate relevant information on financial products and services)

Figure 5.3 Recommendations for enhancing the transparency of financial products and services.

of various bank products and services (bank officials do not provide complete information on the value all products defining only the basic tariff of key activities); (2) absence of an organized and standardized tariff information concerning the financial products of the bank (this makes it difficult for consumers to compare the products of banks among themselves in order to make effective economic decisions); (3) bank staffs offer certain financial products and services based on customer income not their needs (Totolo et al., 2017).

Figure 5.3 gives general recommendations for solving this issue. They allow the regulator to introduce standardized disclosure requirements for financial products and services at a mandatory level for all bank institutions to enable consumers to easily compare key product parameters and make informed choices based on this information.

Since financial products are increasingly complicated, traditional bank products are combined with non-bank products, this leads to the fact that financial products become more difficult for consumers. To avoid distortion of the facts about a financial product by consumers, the bank staff should be familiarized with regulations on the transparency of financial products and services, provide and explain relevant information concerning each product or service of the bank institution, since the consumer needs to fully understand product before buying it and exposing itself to unreasonable risks. The introduction of a standardized consolidated report by regulator will enhance consumer understanding of the nature of the product and will make it easy to compare offers made by different banks before purchasing a bank product or service. It certainly will speed up the choice and allow the consumer to form a complete picture of the full value of the product or service. This report should contain the main conditions of the bank's products and a systematic list of fees to determine the amounts and rates of payment for bank services. It is advisable to delineate a consolidated bank report on financial products and services for individuals in separate sections, namely: products and services in current accounts; products and services for deposits and personal savings (deposit) certificates; cash services; credit products and services; remote banking system "Internet Banking".

5.5 Conclusions

Implementation of Sustainable Development Goals (SDGs) inevitably associated with good practices in public governance and sustained efforts from effective and accountable public institutions. Since global financial crisis has been associated with trust crisis and the loss of financial and other public institutions' reputation public trust in institutions as well as interpersonal trust among citizens has become of important concern for successful reform agenda. It was revealed trust to be considered as multidimensional, heterogeneous and multilevel category contained of three components – interpersonal, institutional and systemic linked by close relationships between them. Multilevel concept for restoring trust in financial system of the country identifying and structuring public trust makes it possible to identify the sources of trust-building on the interpersonal, institutional and systemic levels. Since building effective institutions in the context of SDGs implementation on institutional level requires development of accountable and transparent institutions at all levels, case of ensuring transparency in financial system of Ukraine was presented. In the context of Ukrainian financial system reform originated in the wake of the signature of the EU-Ukraine Association Agreement recommendations for enhancing the transparency of financial products and services will allow the regulator to introduce standardized disclosure requirements for financial products and services at a mandatory level for all bank institutions to enable consumers to easily compare key product parameters and make informed choices based on this information.

Acknowledgement

This work would not have been possible without the financial support of the Ministry of Education and Science of Ukraine. The paper was prepared as part of the Young Scientist Research on the topic "Economic-mathematical modeling of the mechanism for restoring public trust in the financial sector: a guarantee of economic security of Ukraine" (registration number 0117U003924).

References

Akerlof, G. and Shiller, J. R. (2009). Animal Spirits: How Human Psychology Drives the Economy, and Why It Matters for Global Capitalism, Princeton University Press, Princeton. 230 p.

Borum, R. (2010). The science of interpersonal trust. The MITRE Corporation. McLean, VA, 2010.

Buriak, A. (2017). Trust in Banking: Global and Ukrainian Aspects, Financial Institutions and Financial Regulation – New Developments in the European Union and Ukraine, Andreas Horsch, Larysa Sysoyeva – Conference Proceedings. Cuvillier Verlag Gttingen, Germany, pp. 145–162.

Bützer, S., Jordan, C. and Stracca, L. (2013). Macroeconomic imbalances: A question of trust? ECB Working Paper No. 1584. Available at SSRN: https://ssrn.com/abstract=2311266.

Carlin, R., Love, G. and Smith, C. (2017). Measures of interpersonal trust: Evidence on their cross-national validity and reliability based on surveys and experimental data. OECD Statistics Working Papers № 10. OECD Publishing, Paris.

Carney, M. (2013). Rebuilding Trust in Global Banking. Speech at 7th Annual Thomas d'Aquino Lecture on Leadership. Ontario, Feb. 25, 2013.

Comprehensive Program of Ukraine's Financial Sector Development Until 2020. https://bank.gov.ua/control/uk/publish/article?art_id=32802659&cat_id=32893159. Accessed on: 10 October 2016.

Confidence indicators and economic developments: Monthly Bulletin, ECB. Jan. 2013. More than a feeling: confidence, uncertainty and macroeconomic fluctuations: Working paper Series, ECB. Scpt. 2017.

Enhancing Bank Transparency (1998). Basel Committee on Banking Supervision. Available at: https://www.bis.org/publ/bcbs41.pdf (access date October 25, 2017).

Financial Services Results: Edelman Trust Barometer (2018). URL: https://www.edelman.com/trust-barometer.

Fleck, F. and von Lüde, R. (2015). Restoring Trust and Confidence at the Institutional Level by Higher Order Control. The Case of the Formation of the European Banking Union, A Journal on Civilization, BEHEMOTH, № 8, pp. 91–108.

Fukuyama, F. (1995). Trust: The social virtues and the creation of prosperity. New York: Free Press, 1995.

Gillespie, N. and Hurley, R. (2013). Trust and the global financial crisis: handbook of advances in trust research. In: Reinhard Bachmann and Akbar Zaheer (eds.). Cheltenham, Edward Elgar Publishing, 177–203. https://doi.org/10.4337/9780857931382.00019.

Good Practices for Financial Consumer Protection (2017). The World Bank Group, available at: https://openknowledge.worldbank.org/bitstream/ha ndle/10986/28996/122011-PUBLIC-GoodPractices-WebFinal.pdf?seq uence=1&isAllowed=y (access date December 20, 2017).

Gros, D. and Roth, F. (2010). The Financial Crisis and Citizen Trust in the European Central Bank, Working Document No. 334, Centre for European Policy Studies, 26 July.

Guiso, L., Sapienza, P. and Zingales, L. (2008). Trusting the Stock Market. The Journal of Finance, № 63, pp. 2557–2600.

Knack, S. and Keefer, P. (1997). Does Social Capital Have an Economic Payoff? A Cross-Country Investigation. Quarterly Journal of Economics, № 112(4), pp. 1251–1288.

Knack, S. and Zak, P. J. (2002). Building trust: public policy, interpersonal trust and economic development. Supreme Court Economic Review, Vol. 10, pp. 91–107.

La Porta, R., Lopez-de-Silanes, F., Shleifer, A. and Vishny, R. (1997). Legal determinants of external finance, Journal of Finance, № 52, pp. 1131–1150.

Mosch, R. and Prast, H. (2008). Confidence and trust: empirical investigations for the Netherlands and the financial sector. DNB Occasional Studies, Vol. 6, No. 2, pp. 1–65.

OECD (2013). Trust in government, policy effectiveness and the governance agenda, in: Government at a Glance, OECD, OECD Publishing, Paris. http://dx.doi.org/10.1787/gov_glance-2013-6-en.

OECD (2017). Trust and Public Policy: How Better Governance Can Help Rebuild Public Trust, OECD Public Governance Reviews, OECD Publishing, Paris. http://dx.doi.org/10.1787/9789264268920-en.

Putnam Robert, D. (1993). Making democracy work: Civic traditions in modern Italy. NJ: Princeton University Press. Princeton.

Roth, F. (2009). Does too much trust hamper economic growth? Kyklos, Vol. 62, No. 1, pp. 103–128. doi: https://doi.org/10.1111/j.1467-6435.2009.00424.x.

Ruben de Bliek (2013). The development of trust and its relation to economic performance, ICEBI Proceedings, pp. 24–26.

Sapienza, P. and Zingales, L. (2012). A trust crisis, International Review of Finance, Vol. 12(2). pp. 123–131. doi: https://doi.org/10.1111/j.1468-2443.2012.01152.x.

Schatz, R. and Watson, L. A. (2011). Trust Meltdown II, INNOVATION Publishing Ltd, Fribourg.

Simpson, J. A. (2007). Foundations of interpersonal trust: A. W. Kruglanski and E. T. Higgins (Eds.). Social psychology: Handbook of basic principles. New York: Guilford, 2007. 2nd ed. pp. 587–607.

Tonkiss, F. (2009). Trust, confidence and economic crisis, Intereconomics, Vol. 44(4). pp. 196–202, doi: https://doi.org/10.1007/s10272-009-0295-x.

Totolo, E., Gwer, F. and Odero, J. (2017). The price of being banked. Nairobi, Kenya: FSD Kenya, available at: http://fsdkenya.org/publication/the-price-of-being-banked/ (access date February 19, 2018).

United Nations (2016). Compendium of innovative practices in Public Governance and Administration for Sustainable Development, United Nations Publication.

van Esterik-Plasmeijer, Pauline W. J. and Fred van Raaij, W. (2017). Banking system trust, bank trust, and bank loyalty. International Journal of Bank Marketing, Vol. 35, No. 1, pp. 97–111.

Zak, P. J. and Knack S. (2001). Trust and Growth. The Economic Journal, № 111, pp. 295–321.

6

Labor Value Dilemma: Value-Based Concept Application Towards Sustainable Development Goals

**Olexander Zaitsev[1], Shvindina Hanna[2,3,*]
and Medani P. Bhandari[1,4]**

[1]Finance and Entrepreneurship Department, Sumy State University, Ukraine
[2]Department of Management, Sumy State University, Ukraine
[3]Purdue University, USA
[4]Akamai University, USA
E-mail: shvindina.hannah@gmail.com
*Corresponding Author

This chapter deals with the indicator 8.5.1 "Average hourly earnings of female and male employees, by occupation, age and persons with disabilities" for target 8.5 "By 2030, achieve full and productive employment and decent work for all women and men, including for young people and persons with disabilities, and equal pay for work of equal value" for Goal 8 "Promote sustained, inclusive and sustainable economic growth, full and productive employment and decent work for all" (Sachs et al., 2016; Resolution, UN, 2018).

6.1 Introduction

"Decent work and economic growth" were proclaimed as a Goal 8 within SDGs Framework aimed to promote sustained economic growth, higher levels of productivity and technological innovation (United Nations, 2015).

"Goal 8. Promote Sustained, Inclusive and Sustainable Economic Growth, Full and Productive Employment and Decent Work for All
8.1 Sustain per capita economic growth in accordance with national circumstances and, in particular, at least 7 per cent gross domestic product growth per annum in the least developed countries

8.2 Achieve higher levels of economic productivity through diversification, technological upgrading and innovation, including through a focus on high-value added and labor-intensive sectors

8.3 Promote development-oriented policies that support productive activities, decent job creation, entrepreneurship, creativity and innovation, and encourage the formalization and growth of micro-, small- and medium-sized enterprises, including through access to financial services

8.4 Improve progressively, through 2030, global resource efficiency in consumption and production and endeavor to decouple economic growth from environmental degradation, in accordance with the 10-year framework of programs on sustainable consumption and production, with developed countries taking the lead

8.5 By 2030, achieve full and productive employment and decent work for all women and men, including for young people and persons with disabilities, and equal pay for work of equal value

8.6 By 2020, substantially reduce the proportion of youth not in employment, education or training

8.7 Take immediate and effective measures to eradicate forced labor, end modern slavery and human trafficking and secure the prohibition and elimination of the worst forms of child labor, including recruitment and use of child soldiers, and by 2025 end child labor in all its forms

8.8 Protect labor rights and promote safe and secure working environments for all workers, including migrant workers, in particular women migrants, and those in precarious employment

8.9 By 2030, devise and implement policies to promote sustainable tourism that creates jobs and promotes local culture and products

8.10 Strengthen the capacity of domestic financial institutions to encourage and expand access to banking, insurance and financial services for all

8.a Increase Aid for Trade support for developing countries, in particular least developed countries, including through the Enhanced Integrated Framework for Trade-Related Technical Assistance to Least Developed Countries

8.b By 2020, develop and operationalize a global strategy for youth employment and implement the Global Jobs Pact of the International Labor Organization".

Source: United Nations 2015
https://sustainabledevelopment.un.org/post2015/transformingourworld

Despite the facts that over the past 25 years the poverty declined (Beltekian and Ortiz-Ospina, 2018), and labour productivity since 2001 has consistently grown (Decent work…ILO Report, 2018), the disproportions in the development remain. For instance, the growth was faster in middle-income economies than in high-income economies (ILO, 2018).

6.2 Literature Review

The disparities in regional development in EU and globally became one of the urgent topics in a sphere of labor productivity assessment and social policy making. The interrelations between hourly costs, hourly salary in different parts of the countries led to internal migration, e.g. from rural areas to cities (Lipton, 1980), and disproportions hourly salaries and standards of living between countries became one of the causes of to international migration (Gorter et al., 2018), as waves of migration can be perceived as typical movement from one area to another seeking better social an economic prospects. At the same time, the disproportions in labour costs remain one of the most interesting phenomena to investigate considering the recent changes in education and knowledge production, and most importantly in labour productivity due to innovations. The access to technologies and labour productivity are interrelated (Jones, 2016), and therefore, the labour costs differ from region to region. The study of Bosetti and his co-authors (Bosetti et al., 2015) is concentrated on high-skilled foreigners' contributions into innovative capacity of the host countries. Meanwhile Filippetti and Peyrache (2015) consider technology gap in Europe as the main concern for policy-makers. The shifts in economic development (and therefore in labour costs) can be explained by difference in R&D activities and technological capabilities of different regions. However, given research is aimed to detect the main disparities and disproportions in economic development in terms of labour value difference in different regions, and to offer some new insights for understanding of 'equal value' related to Goal 8 of SDG 2030 Agenda.

6.3 Progress and Challenges in Sustainable Development Goal 8

According to the Sustainable Development Goals Report (The Sustainable Development Goals Report, 2018), labour productivity has increased globally in 2018, but at the same time there are still the problems with labour market inequality, safety and security of work environment, inclusiveness and "decent work for all", as it is mentioned by 8th SDG goal. There are some serious improvements, which should be mentioned as follows:

- Labour productivity at the global level, presented as output produced per employed person grew by 2.1 per cent in 2017 (measured in constant 2005 US dollars), this is the fastest growth observed since 2010.
- The global unemployment rate decline (from 6.4 per cent in 2000 to 5.6 per cent in 2017).

At the same time, 'informality worldwide' stays as a trend, as more than 61 per cent of all workers were engaged in the informal economy (SDG, 2018) globally, and it should be noted that the highest rate of informal employment is in emerging and developing countries.

The gender inequality in earnings is still pervasive: the hourly wages of men are, on average, higher than those of women, with a median pay gap of 12.5 per cent (89 per cent of countries out from 45 observed countries, ILO 2018). The same lag exists for having an account at a bank – women are behind men in this regard, as well as youth are behind adults (especially in low-income countries). *"The gender gap in pay has narrowed since 1980, but it has remained relatively stable over the past 15 years or so. In 2018, women earned 85% of what men earned, according to a Pew Research Center analysis of median hourly earnings of both full- and part-time workers in the United States. Based on this estimate, it would take an extra 39 days of work for women to earn what men did in 2018. By comparison, the Census Bureau found that, in 2017, full-time, year-round working women earned 80% of what their male counterparts earned. The 2018 wage gap was somewhat smaller for adults ages 25 to 34 than for all workers 16 and older, our analysis found. Women ages 25 to 34 earned 89 cents for every dollar a man in the same age group earned"* (Graf 2019-PEW).

It is still big youth unemployment rate, 13 per cent in 2017 which means that youth is three times more probably to be unemployed. As it's observed, the informality problem, labour market inequality remain to be a challenge towards sustained and inclusive economic growth. The difference in salary between countries is huge, especially if to compare proportions of low-wage earners in total number of employees, the best positions are belonged to Sweden and the highest gap in earnings is observed in Latvia (see Figure 6.1).

The disproportions and disparities in regional development between different EU and non-EU countries are obvious when labor costs are considered. It can be explained by number of hours spent for work, difference in technologies, productivity of the country in terms of national competitiveness level, and specialty of the country.

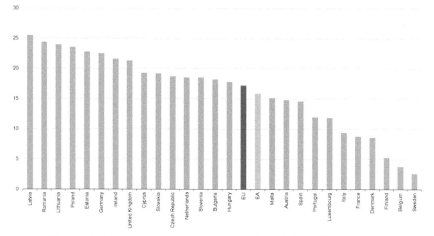

Figure 6.1 Low-wage earners as a proportion of all employees, 2014.
Source: EUROSTAT Database.

The disproportions and differences may be presented through the efficiency indexes, for example, the global competitiveness index (GCI) or labour productivity can be taken as an indicator which can be evaluated as efficiency of resources using and quality of the work. Not only the output of the labour is measured to perform the efficiency, but the value added per hour worked. In this case if we compare extreme situations of low-earners disproportions, or low average salary (Latvia) with the one of most favorable regions in terms of living standards and high average wage (Luxemburg), we may see the explanation: (i) the difference in gross value added (GVA), (ii) specialization of the region, for instance, the highest share of financial and insurance activities in total gross value added (27.5%) is registered in Luxembourg, while in Latvia the only leading sphere in terms of GVA is real estate activities.

As we may assume the specialty of the country, which we understand as a dominance of certain sectors, can explain the differences and disproportions in the development, such for instance, Romania has the highest share of employment in agriculture and the second highest low-earners proportion in total employment. The link between GVA, GDP per capita and labour market inequalities should be further investigated, given previous findings in this field, considering the labour productivity as an indicator influenced by multiple interrelated economic, political, institutional factors which act simultaneously, and which dynamic is unpredictable.

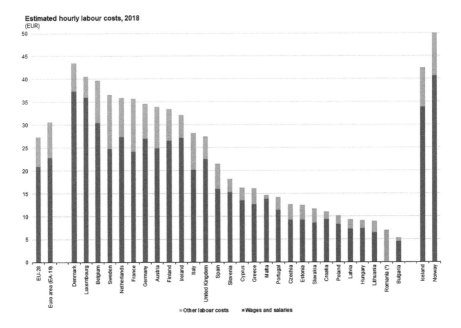

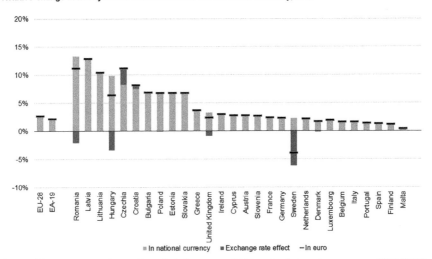

Figure 6.2   Data on estimated hourly labor costs, 2018 and relative change costs 2018/2017.
Source: EUROSTAT Database.

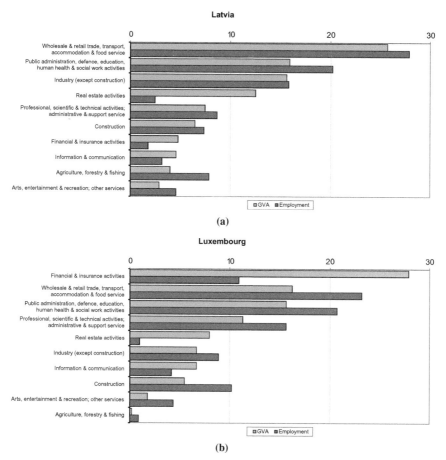

Figure 6.3 Gross Value Added (GVA) and Employment by the economic activity as % of total, in (a) Latvia, and in (b) Luxemburg.

Source: EUROSTAT Database.

6.4 Development of the Labor Value Based on Civil Society Evolution

"Civil society is the plethora of private, nonprofit, and nongovernmental organizations that have emerged in recent decades in virtually every corner of the world to provide vehicles through which citizens can exercise individual initiative in the private pursuit of public purposes. If representative government was the great social invention of the eighteenth century, and bureaucracy – both

public and private – of the nineteenth, it is organized, private, voluntary activity, the proliferation of civil society organizations that may turn out, despite earlier origins, to represent the great social innovation of the twentieth century".

(Salamon and Anheier, 1996:1 as cited in
Bhandari and Oli, 2018:71)

The origin and evolution of the category "civil society" has four scientific and at the same time historic periods.

1st Period. There appears a problem to define a civil society because the civil society has three different aspects – social, economic and political. Since the greatest of the Greeks Aristotle and up to the Englishman John Locke (1632–1704) only two spheres, social and political, have been dealt with in the unity and considered to be parts of the state system. There was no civil society as a separate category. The society in the unity of political and social forms was regarded as the category "state". Societies, according to their structure, were, firstly, political societies and, secondly, social societies, and such understanding prevailed until the middle of the XVII century when John Locke wrote his "Second Treatise of Government" (1662). In one of the chapters titled 1890–1907 "Political or Civil Society", J. Locke said that a society is a more complicated system, it differs from a patrimonial, tribal, feudal or monarchic political union, moreover, it differs significantly from a family. One of the aspects of a state was called a civil society by John Locke. Besides, according to J. Locke, a civil society is incompatible with the absolute monarchy. Besides it is a "body" to build a political state. Locke also advocated governmental differentiation of powers and believed that revolution was not only a right but an obligation in some circumstances. These ideas would come to have a profound influence on the Declaration of Independence and the Constitution of the United States (Locke and Kelly, 1991).

A century later the terminology changed. The work by the Scottish Adam Ferguson (1723–1816) "An Essay on the History of Civil Society" (1767) marked the gap between the political and social spheres of a state. Approximately at the same time James Madison (1751–1836) who was an American statesman and the fourth President of the United States emphasized the role of a civil society as a counterbalance to a state in his articles in the magazine "Federalist". He thought different groups of the society that have versatile interests to be the guarantee from the majority tyranny. The concept of the civil society was thoroughly elaborated in the works of scientists and enlighteners of XVII–XIX centuries. They included the English philosopher Thomas

Hobbes (1588–1679), the Dutch philosopher Baruch Spinoza (1632–1677), the French writer Charles-Louis de Montesquieu (1689–1755), the French philosopher Jean-Jacques Rousseau (1712–1778), the German philosophers Immanuel Kant (1724–1804) and Georg Wilhelm Friedrich Hegel (1770–1831) (Sartori, 1987, 1991; Rienner, 1989; Linz, 1978; Fromm, 1942; Rawls, 1971; Bhandari and Oli, 2018).

Summary of 1st Period. There appear and evolve new principles of state system that should replace a feudal-monarchic state. The main idea of these researches consists in the fact that an old state "should be replaced with a new state system", and the "new state system" is described as the one that is bound to be a civil society. This is the period of the formation of ideas about a civil society for the future. Such a civil society is a unified integrity of all society aspects. Such a civil society is an inseparable unity of the social, political and economic fields of the state. Such a civil society is a just social society from the very beginning, and both politics and economy operate according to this principle of justice. The definition of a civil society in the field of Period 1 is exclusively societal, initially social, but without any economic specificity. A civil society is a union of individuals, individual atoms, united by a society. But at the same time, a state does not prevail over a civil society, since the state and the civil society are the same thing.

2nd Period. A civil society, according to Adam Smith (1723–1790), is a developed capitalist society in its maturity. The civil society, according to Adam Smith, is the goal of development, a type for a future state system based on new liberal principles. A. Smith suggested a logical scheme that explained the work of the free market based on inner economic mechanisms, but not on the external political government. Up to now this approach has been the basis of the economic education. The free market forms a new society. This is the society in which the leading role is played by economic relations based on human labor. A. Smith saw the new civil society built on the principles considering the laws of labor value. These laws act disregarding the desires of people, that is why they could not be neglected, but they should be considered in the life of the society.

The Scottish A. Smith as well as the French economist Jean-Baptiste Say (1767–1832) and the Swiss Jean Charles Léonard de Sismondi (1773–1842) supporting his doctrine in a quarter of the century, the Englishman David Ricardo (1772–1823) successfully developing his doctrine forty years later and many other supporters are considered the founders and developers of the economic trend that is called the classical political economy now.

At that period the classical notions of a civil society were based on the ideas of the free market economy that, in fact, began to act in the economic relations of some European countries (Netherlands, Great Britain, France and Denmark). However, speaking about the new societal system, the new societal system of a state was often mentioned (as in A. Smith's works) disregarding the term "civil society" itself. Its essential characteristics, in political economy classics' opinion, is the society with the distinct private activity, different from the family and existing apart from the state. A. Smith, D. Ricardo and a lot of their followers equated the new capitalist relations with the new civil society which Hobbes and Spinoza, Locke and Ferguson, Madison, Montesquieu and others implied.

Summary of 2nd Period. Here the emerging economic aspect was substantiated as one of the three aspects of a civil society.

3rd Period. The well-known representative of the German philosophical school G. W. F. Hegel based his notion of a civil society on two fundamental principles: individuals pursue only their private interests, that is, they behave like isolated atoms, and, herewith, the social connection among them is formed when everybody depends on each other. Here "the system of atoms" is an accordance with the doctrines of Hobbes and Spinoza, Locke and Ferguson, Madison and Montesquieu, etc. On the other hand, Hegel realizes the social connection through the economic interactions of individuals, like A. Smith and D. Ricardo. The civil society as an economic structure represents the system of market relations. The civil society in G. Hegel's interpretation is the system of needs mediated by labor which rests on the dominance of private property and universal formal equality of people. G. Hegel dealt with the civil society as the form of people's community which is generated by the constant connection between the family and the state and which provides the viability of the society and application of civil rights of every of its members (Hegel, 1990).

Hegel suggests such logic of the human society development sequence. The social being exists: (1) as immediate, i.e. the natural interaction of people that is a family; (2) as attitudes of certain people and independent individuals to each other, (and the family is one person) within some formal community, and this is a civil society; (3) as a state when it is a combination of principles of a family and a civil society, §517 (Hegel, 1830). Certainly, it is the philosophical vision of the statehood evolution by G. W. F. Hegel, the representative of the German philosophy of the beginning of the XIX century. It should be noted that it is not the definition of a state (Item 3) in this Hegel's construction but the place of the civil society with regard to the state (Item 2) that matters for us. The civil society in its historical evolution appears and functions earlier than the state appears. G. Hegel gives the following definition of a civil society. "A civil society is a kind of universal, stable social connection that unites the interests of members of society. The developed totality of this connection is the state as a civil society", §523 (Hegel, 1830).

The civil society in Hegel's interpretation is historically a forming element of building a state. It should be noted again: "The developed totality of this connection is the state as a civil society", §523 (Hegel, 1830). Therefore, such a regularity is evident. If a state suppresses its own civil society inside itself, it will destroy itself with such actions as it destroys its own foundation. The vivid example of the self-destruction in the modern history is the internal policy of the Russian Federation at the end of the twenties of the XXI century. The disintegration of the USSR at the end of the XX century has also one of the reasons – the struggle of the state ideology against the self-consciousness of the civil society that is wider than the official state ideology. To suppress and struggle against the civil society existing in the state is the same thing as to cut off the twig on which this civil state "has been settled" and "has been developing". The example of contradictions between the civil society and the old-fashioned statehood is the Civil war in the USA (1861–1865) between the industrial North and the slave-holding South. By the beginning of the 60s of the XIX century the civil society of the USA had outgrown the slavery foundations in its development and, as a result, destroyed such a state.

Karl Marx (1818–1883) "turned over" the structure of Hegel's civil society model and put labor, production and exchange into the foundation of the civil society functioning. K. Marx did it in this way because he followed the principles of the economic doctrine of the classical political economy by A. Smith and D. Ricardo. Relying on such an economic model, K. Marx developed the new opinion concerning the place and role of a civil society in the system of public and state relations. Marx's point of view is formulated in the following excerption.

Speaking briefly about the general state order, Marx criticizes the statement that the universal state order must hold together the individual self-seeking atoms. The members of the civil society are not atoms. It is the natural necessity, the essential human properties and interest that hold the members of a civil society together; a civil, not political life is their real tie. Thus, it is not the state that holds the atoms of civil society together, but the fact that they are atoms only in imagination, and in reality they are living beings tremendously different from atoms as they are human beings. It is the political superstition that can still imagine today that the civil life must be held together by the state, whereas in reality, on the contrary, the state is held together by the civil life (Marx, 1955).

Summary of 3rd Period. According to Hegel, on the one hand, a civil society is a constructive element of developing a state and its functioning, i.e. a civil society is a part of a state. On the other hand, a civil society is a system of economic (market) relations. From the economic point of view, a civil society is a state. That is why, a civil society has an immanent economic constituent, it is the basis of state development and the factor of its vital activity, at the same time it is, in fact, a state itself. According to Marx, relations (according to G. Hegel – connections) among people are based on the material life relations (property relations, production relations, exchange relations, distribution relations). Hegel called the combination of such life relations, using the example of English and French writers of the XVIII century, "a civil society". Thus (according to Marx), the essence of a civil society and mechanisms of governing it should be viewed through the economic relations.

4th Period. The modern interpretation of the term "civil society" in Ukraine and the Russian Federation will be considered. In Ukraine the term "civil society" is used by specialists from the sphere of law, political science and sociology. The typical definition of a civil state from the textbook that was most popular in Ukraine at the beginning of 2000's is given as "the civil society is the system of interrelations among individuals and their groups in which the individual and collective interests are realized. It is based on the autonomous principles protected from the excessive intervention on the part of the state, on the freedom of self-realization, the pluralism in all the spheres of public life and the priority of human rights" (Volynka, 2003). But in the next sentence of the text there are words refuting this definition immediately. The refutation says: "Certainly, such relations between a state and a civil society appear when both a state and a society are on the high level of economic, political, cultural and spiritual development" (Volynka, 2003). It is interesting to know how one is to define the height of this level and in which modern state such relations are on the high level of economic, political, cultural and spiritual development at present. The given definition only declares something desirable abstracted from reality. The things are not better with the understanding of the civil society essence in the Russian Federation as well. We will illustrate it with the definitions of the two famous Russian scientists, K. S. Gadzhiyev and Z. T. Holenkova.

In Gadzhiyev's opinion, a civil society is the system supporting the life activity of the social, cultural and spiritual spheres, their reproduction and transmission from generation to generation, the system of autonomous public institutions and relations independent of the state that are to provide the conditions for the self-realization of separate individuals and groups, the realization of private interests and needs no matter whether they are individual or collective (Gadzhiyev, 1991). In the definition there is not a hint at the economic aspect of the civil society. The whole economy is neither "in the hands" nor "under the control" of the civil society. But if the economy is neither "in the hands" nor "under the control" of the civil society, which instruments and mechanisms does the civil society support systemically the life activity of the social, cultural and spiritual spheres with? In K. S. Gadzhiyev's interpretation a civil society is fully separated from the state in political and economic aspects and allegedly autonomous from the state. So, in Holenkova's opinion, a civil society is a specific combination of social communications and social connections, institutions and values, whose main subjects are a citizen with his/her civil rights and civil (neither political nor state) organizations: associations, unions, public movements and civil institutions (Holenkova, 1997).

Both Holenkova and Gadzhiyev think a civil society to exist not into a state, but separately from it. On the other hand, a civil society is a specific combination of social communications and social connections, institutions and values, but not political or state ones. The economic social communications are not mentioned at all. The definition of a civil state is put into the plane of moral and subjective (in Holenkova's definition) and spiritual and cultural (in Gadzhiyev's definition) values. Such a notion of a civil society was characteristic for the medieval Europe before the works of Hobbes and Spinoza, Locke and Ferguson, Madison and Montesquieu.

Summary of the 4th Period. In such understanding the term "civil society" is confined to general public and and to a multitude of public organizations autonomous from the state, it makes this view senseless.

Thus, we can generalize the notion of a civil society as the system of the society vital life aiming at the state regulatory mechanism, but not as the autonomous system in relation to the state power, expressing private individual, group, corporative interests of citizens, regulating and protecting these private interests. Hypothetical analogy is appropriate to explain further. For instance, some building materials, construction machinery and tools are necessary to build a house, qualified workers, a brigade of builders, are also needed. If the house that is being built may be hypothetically called a state, the building materials – the elements from which the state is "constructed", the brigade of builders' activity with the use of the construction machinery and tools is nothing but the analogy of the civil society. Moreover, such an activity never stops! In other words, a civil society is an immanent process, an activity which "creates" the state by means of public and state activities of independent people and families. The real productive relations, or economic relations that are the same, "build" the state, not vice versa. The economic relations are public relations, that is why they are peculiar of a civil society. It is in such a way how Hegel understood it. This was the way in which the developers of the classical economy who developed economic theories in the second half of the XIX century understood it. From this perspective, a civil society was viewed as a constantly proceeding social process of building the statehood from the bottom to top according to the scheme: social individuals and/or families (households) → civil society → state. But not vice versa, this not creation of the state at the beginning which the state in its turn as the organizer and the controller creates the civil society.

> *"The advantage that NGO have over the public sector is the freedom from fixed administrative procedures or standard operating procedures. This led Donor enthusiasm to a proliferation of NGO, many of them not all that motivated by altruism [because the motives of NGO leaders] may be exactly the same as those of a for-profit firm – requiring the same monitoring and care in contract enforcement".*

(World Bank, 1995)

6.5 Development of Classical Economy Towards the Civil Economy

The need to produce commodities and the production means, the subsequent exchange of these commodities has been taking place, but not because these processes were initially centralized and purposefully organized by a state. Rather the opposite. The production of goods and their exchange appeared long before the appearance of the state as a political form of society organization on a certain territory. Before the appearance of the state, there had already been such social forms as family (household), clan, tribe, union of tribes, and they had their own economic forms. The formation and development of economic forms of social development had existed before the emergence of the state. It is the development of economic forms of relations among people that caused a state to appear.

The forms of commodity production, benefits, consumer values to maintain the lives of society members, i.e. the economic forms of arrangement and organization of social life, precede the state historically. Therefore, they are the basis on which the political, legal, religious, artistic and philosophical forms of social interaction develop, and generally they are the ideological forms of the state, by means of which people recognize themselves as citizens of their society-state.

We assume that a civil society was "born" out of the vital activity of the human society. Then, in its development, the civil society is the preceding basis in relations to the state. These are such civil relations that build, change and strengthen the state. The civil society is inseparable from the state. The civil society does not separate itself from the state but forms it actively, according to its values and interests. Such a civil society has an economic constituent. The economic relations were said to be public relations. Thus, a civil society has a specific civil economy of its own and the basis for the civil society.

For the last three hundred years, all the classical political economy has been targeted on the search for such an economic state model that would correspond with the goals and interests of the civil society. That is why the classical economic theory, from A. Smith to K. Marx, can be understood as a scientific way of elaborating the economic theory for the civil society. The task of the civil economy is finding out the laws and regularities in economy which are independent of human desires and tricks, and if there are such ones, then to build a society according to them and use these regularities for the benefit of people. For instance, all the objects and people on the Earth that move or that are in the state of rest are affected by the force of gravity. If the gravity effect is neglected, the buildings and facilities will break down, airplanes and people will fall. The entire world will be an ever-lasting catastrophe. So are the things in economy. If the law of value is neglected, all the economy will have periodic catastrophes that are called economic crises.

Marx's elaboration of economic ideas of A. Smith and D. Ricardo within the framework of G. Hegel's philosophical views caused him to perform labor as the measure of social value. The universal economic connection among the members of the society is labor that is spent on the production of goods, namely, its quantitative characteristics with respect to some goods. The amount of labor spent on the commodity production in the developed capitalist society influences the quantitative proportions during the exchange of goods and services. Moreover, it influences them so strongly that it acts as an economic law. In the economic theory that studies it this is the law of value. The use of value indicators in economy alongside with money indicators is the development of the economic thought towards the civil economy.

The measurement of value which is not measured in money but has another indicator will be discussed below. Nowadays value indicators have money measure, and the money indicator of value results from selling commodities or after payment of the delivered services. If the fact of selling a commodity or a service took place it means that that their value at that time in a certain place is equal to the money sum that was paid for them. It is the essence of forming (the appearance of) the value measure.

Let us have a more attentive look. When the commodities and services are sold there are people near them – sellers and buyers. It is they who agree upon the value quantity in the form of money. Therefore, in the first place, money value is the result of agreement among people at the moment of purchase/sale of commodities or services. No doubt, this agreement is influenced by a lot of other factors but, as a result, it is the agreement between a seller and a buyer that is the factor determining the money value quantity finally. That is why, we will call such a value of commodities and services the subjective value. However, there is another assessment of value of commodities and services that does not depend on the human agreement, that does not depend on the convention among people, that is abstracted from human desires. Let us call such a value the objective value.

Such value assessment was mentioned by William Petty (1623–1687) (Martin et al., 2009). He distinguished the so-called "political" and "natural prices". Under "political prices" Petty understood constantly changing market prices, and under "natural" ones a certain amount of labor is meant that is spent on producing commodities. Then Petty's term "political prices" is a subjective value, and his term "natural prices" is an objective value, in our formulation. According to Faccarello (1999), at the same time with Petty and independently of him French scientist Pierre Boisguillebert (1646–1714) differentiated a market price and "true price" in his works. He thought that the true value is natural, determined by labor spent on the commodity production while "market prices" are random. Boisguillebert determined its quantity by labor time. "A true value" was assumed as a basis of the balanced exchange by Boisguillebert. So, we may resume that Boisguillebert's "market price" is a subjective value and his "true value" is characteristic of the objective value.

Both Adam Smith and David Ricardo considered that commodities form their exchange value from two sources – their scarcity and amount of labor that are necessary for their assumption. There are commodities whose value is determined exclusively by their scarcity (sculptures, pictures, rare books, coins). The value of these commodities is not determined by the amount of labor spent on their production, it changes depending on tastes, desires and financial assets of a consumer. But their amount is slight in the general mass of commodities. According to Smith and Ricardo the exchange value of the major mass of commodities is determined by the amount of labor. It should be mentioned that Smith and Ricardo have another point of view. In their opinion, during the exchange the price of commodities is formed by two factors: both by scarcity, i.e. demand and supply, and amount of labor spent on their production. In other words, subjective value and objective value participate together in forming the money price.

Marx in his book "Capital" (Marx, 1867, 1887) presented the value that is measured by the amount of labor (objective value – our explanation) to be a basis, a core of economic processes. Hence, Marx stated that market prices (subjective value – our explanation) increase or recede as a result of being influenced by the objective value. Afterwards the followers of K. Marx's economic doctrine began to call his economic concept the labor theory of value. The major question an answer to which was sought both by opponents and supporters of the labor theory of value remains and has always been the following one. How should the labor value of commodity be measured quantitatively?

6.6 Labor Input as an Indicator of Value Measurement

The overwhelming majority of calculations in different spheres of economy are connected with calculations of value indicators. Currently the value assessment of products, commodities, material objects and services are measured by money indicators. Now the term "value" denotes the indicator in form of the sum (quantity) of money, and no other numeric indicators are implied with respect to this term in economy and finance. However, together with money assessment which has historically evolved as the one that is expressed quantitatively, i.e. numerically, for four centuries or so the economic science has been paying attention to the notion of value as the indicator measured by money. Starting with the works of A. Smith, D. Ricardo, the basis of value of any object made by people, anything produced by them is labor that is regarded to be human efforts necessary to produce a certain object or thing. The value interpreted as human labor costs underlies the assessment of commodities and services in the theory that is called classical political economy. The economic doctrine of A. Smith and D. Ricardo served the basis of K. Marx's economic theory. Further, the economic literature began to call the researches of K. Marx and his supporters the theory of labor cost. At the same time, neither A. Smith, D. Ricardo, K. Marx nor any of their supporters suggested the indicator to measure spent labor in the form of the numeric indicator measurer.

In Volume 1 of "Capital" K. Marx defines that labor costs are measured by the amount of working time hours. Relying on the surface interpretation of the phrase "working time hours" (reference), the followers of the classical political economy think that value is measured by labor costs in astronomic hours, i.e. the number of hours which are counted by mechanical or electronic clocks. But the amount of the astronomic time during which a worker (an employee, a manager) works does not indicate labor costs. See more details about it in (Zaitsev, 2004, 2010, 2016). There is another functioning mechanism which does result in the numeric indicator of the amount of spent labor. The observation of a worker while he/she is performing some working functions provides a range of reliable indicators characterizing the labor amount. They are described in more detail in (Zaitsev, 2004). In order that we might identify the indicator describing the quantity of labor we choose one of them, namely, the heart rate or pulse rate (which is the same) of a worker for a certain period of his work, as a rule, per minute. The observations show that the more labor a worker (an employee, a manager) puts in, the more beats his pulse makes. Such a directly proportional dependence of the number of pulse beats on the amount of labor has been fixed many times, for instance, (Vinogradoff, 1969; Scherrer, 1967, 1973). The concrete types of labor can be divided into groups according to the workload, using the pulse rate as an indicator. In 1963 in England Brown and Growden investigated the outcomes of industrial workers connected with Slough Industrial Health Service and published such data in "Slough Scales" – dependence of pulse rate (heartbeats per minute) on workload: an easy work – 60–100, moderate work – 100–125, hard work – 125–150, very hard work – 150–175 heartbeats per minute (Vinogradoff, 1969). In 1967 a group of scientists from France, Switzerland and Belgium published a book "Physiologie du travail (ergonomie)" (Scherrer, 1967). In this book Monod and Pottier give such a classification of works according to heart rate: very easy work – up to 75, easy work – 75–100, moderate work – 100–125, hard work – 125–150, very hard – 150–175, extremely hard work – more than 175 heartbeats per minute (Scherrer, 1973). In the investigations of the Soviet period (1981) there are such data in (Shybanoff, 1983): very easy work – 70–80 heartbeats per minute, easy work – 80–90, moderate work – 90–100, medium work – 100–125, hard work – 125–150. We choose the easy work as the unit of labor inputs which is characterized by 75 pulse beats per minute, i.e. an hour of labor inputs is equal to 4500 pulse beats. The figure 45000 is calculated as multiplying 75 pulse beats per minute by 60 minutes. So, we can put it down: 1 hour of labor inputs = 1 hour of value = 4500 pulse beats.

The unit we have chosen to measure labor value has got such a dimension: 1 hour of value is equal to 4500 pulse beats of a workman (a worker, an employee, a manager). Then, correspondingly, 2 hours of value are equal to 9000 pulsebeats, 3 hours of value are equal to 13500 pulsebeats, 8 hours of value are equal to 36000 pulsebeats of a workman. Thus, if the indicator of the total number of pulse beats per working day, for instance, on an eight-hour work day makes in total 63000 pulse beats for a particular worker, then we have 14 value hours (63000/4500 = 14). It means that for 8 astronomic working hours a worker (an employee, a manager) has transmitted 14 value hours on the labor product (commodity or service) or the labor of such a worker has produced the value amounting 14 value hours for 8 astronomic hours. Such a "discrepancy" between the number of astronomic working hours and the number of value hours of work is caused by the fact that such worker's quantity of labor takes into consideration his pulse rate that makes, on average, more than 130 pulse beats per minute. i.e. it was a hard work in comparison with the easy one with 75 pulse beats per minute. That is why, such a work has produced higher value in comparison with an easy work for the same period of work during the astronomic time. Such indicators have been widely used in measuring the sport work and characterizing the quantity of labor in space medicine for a long time (Baevskyi, 2003; Volkov et al., 2003). As a result, a new economic indicator is suggested that characterizes the quantity of spent human labor or, in other words, an indicator of the value that differs from the traditional money value: 1 hour of value = 4500 pulse beats of a working individual or 1 value hour = 4500 pulse beats of a workman (a worker, an employee).

6.7 New Value Indicator as a Tool Measure for Goal 8 Progress

This subchapter is devoted to "Global indicator framework for the Sustainable Development Goals and targets of the 2030 Agenda for Sustainable Development". Let us consider "Goal 8. Promote sustained, inclusive and sustainable economic growth, full and productive employment and decent work for all". The indicators presented in 8.5 "By 2030, achieve full and productive employment and decent work for all women and men, including for young people and persons with disabilities, and equal pay for work of equal value" are in the correspondence with Indicator 8.5.1. "Average hourly earnings of female and male employees, by occupation, age and persons with disabilities" (Resolution, UN, 2018).

Here are some critical remarks on this indicator. It is a purely monetary indicator. It shows how many banknotes a workman gets for an hour of his work. This hour is astronomic, and it is measured in hours in a day. There are drawbacks of the indicator "Average hourly earnings of female and male . . . ". The indicator can show growth, but, in fact, it can be decline. The reasons are as follows.

Firstly. Let the indicator of average hourly earnings has grown. It is defined in the following way. Average hourly earnings are being compared with the preceding average earnings, for instance, last January and this January. However, it is not conspicuous in the indicator at whose expense the indicator has grown. It could increase either because of the labor productivity or increasing the working hours, for instance, from 8 to 14 hours per day or depending on the growing work intensity, or of three variants together, but in different proportions.

Secondly. The growth of this indicator shows the growth of incomes for the receiver of earnings in one case only, when the inflation rate does not rise. In fact, economies of all the states have inflation. If inflation grows faster than the earnings do, the growth of earnings in money indicators is a fake.

Thirdly. The correction of this indicator to inflation has its drawbacks. Every state has its methodology of determining inflation and its own list of commodities and services to calculate the inflation rate.

We offer another indicator for 8.5.1. All the above-mentioned drawbacks are removed if the new indicator is comparison of two new value sub-indicators. Let us call the new indicator "Average value earnings . . . ". The first new value sub-indicator shows how many banknotes are paid to a workman for the unit of working time put in by him or, which is the same, how many banknotes are paid to a workman for his one value hour. (It should be mentioned that the working time and the time of work are different indicators. The working time is measured in value hours, and the time of work is measured in diurnal hours). Let us call it an earning sub-indicator. The second new sub-indicator shows how many banknotes are paid for goods at the market in calculating per hour of value. Let us call it a price sub-indicator. After that the first earning sub-indicator is compared with the second price sub-indicator, and that is the new indicator "Average value earnings . . . ". If the result of dividing the first sub-indicator by the second one has numerically increased, "Average earnings . . . " correspondingly grows, but if the result of their division is numerically falling, then the indicator "Average . . . earnings of female and male employees . . . " falls.

The indicator "Average value (highlighted by the author) earnings of female and male employees, by occupation, age and persons with disabilities" has no above-mentioned drawbacks that are pertaining in Indicator 8.5.1 "Average hourly (highlighted by the author) earnings of female and male employees, by occupation, age and persons with disabilities" accepted in the UNO.

There are some difficulties existing in Ukraine and in other countries, perhaps. (1) There is some statistics of the time of work in hours but there is no subdivision of the time of work for female and male employees, as well as for persons with disabilities. (2) There is some statistics, as far as the earning is concerned, however there is no such classification, as for female and male employees and persons with disabilities. (3) There are price expenditures according to consumer goods, but there is no subdivision according to female and male employees. However, this problem is easily solved, namely, all the purchases of food products are conventionally divided depending on the proportion of men and women, all the public payments – in the same proportion. The manufacturing products should be divided depending on the gender destination for men and women.

6.8 Conclusions

> *"The End of our Foundation is the knowledge of Causes, and secret motions of things; and the enlarging of the bounds of Human Empire, to the effecting of all things possible".*
>
> (Francis Bacon, The New Atlantis)

United Nations have been playing very important role to overcome the world problem from its inception. *"The United Nations was founded in 1945 with the mission to maintain world peace, develop good relations between countries, promote cooperation in solving the world's problems, and encourage a respect for human rights. It provides the nations of the world a forum to balance their national interests with the interests of the global whole"* (PBS 2014). With the main goals of (1) Maintain international peace and security; (2) Develop friendly relations among nations; (3) Achieve international cooperation in solving international problems; and (4) Be a center for harmonizing the actions of nations in the attainment of these common ends (UN, 2009). Its roles and responsibilities have been criticized because it can only execute its program through governments. However, it has very important role on peace and security, health, economic growth of developing world, human, women, child rights and since 1972 environmental management. In relation to labor related issues UN agency ILO is playing important role about the equality of labor related issues among men and women labor. However, there is very big difference between men and women's labor payments, both in developed and developing nations. This chapter tried to present some of basic facts of such variations.

In 2015 (United Nations General Assembly), in transforming our world: the 2030 agenda for sustainable development adopted 17 goals and 169 targets among them goal 8, deals with the "Decent work and economic growth" were proclaimed as a Goal 8 within SDGs Framework aimed to promote sustained economic growth, higher levels of productivity and technological innovation (United Nations 2015). This goal also related to Goal 1 – No poverty; Goal 2 – Zero hunger; Goal 4 – Quality education; Goal 10 – Reduced inequalities; Goal 14 – Life below water and Goal 16 – Peace and justice; strong institutions. The achievement of the goals depends on how the governments are or will implement these goals and targets in their national plans and how they execute. This chapter also discussed about the role of civil society organizations related labor issues with historical account. Civil societies are playing critical role in minimizing the gender gap in the labor force, however, the scenario so far does not show significant development. It is essential to empower women and marginalized population to bring to the equality. There are many studies on Labor Value Dilemma: Value-Based Concept (Blau and Kahn, 2017; Goldin, 2014; Goldin and Katz, 2016; Lundborg et al., 2017; Blau et al., 2017; Neumark et al., 1996; Waldfogel, 1998; Olivetti and Petrongolo, 2017); and there are also many initiatives, however, the scenario shows miles to go to achieve the desired outcome.

References

Baevskyi, R. (2003). The concept of physiological norm and health criterions [Kontseptsyia fyzyolohycheskoi normy i kryteryy zdorovia]. *Russian psychological journal named after I. M. Sechenov [Rossyiskyi fyzyolohycheskyi zhurnal ymeny Y. M. Sechenova]*, 89(4), 473–478.

Beltekian, D. and Ortiz-Ospina, E. (2018). Extreme poverty is falling: How is poverty changing for higher poverty lines? Our World in Data, March 05, 2018. Retrieved from https://ourworldindata.org/poverty-at-higher-poverty-lines.

Bhandari, Medani P. (2018). Green Web-II: Standards and Perspectives from the IUCN, Published, sold and distributed by: River Publishers, Denmark/the Netherlands. ISBN: 978-87-70220-12-5 (Hardback) 978-87-70220-11-8 (eBook).

Bhandari, Medani P. and Oli, Krishna P. (2018). 'The changing roles and impacts of civil society/NGOs in Nepal, *Civil Society in the Global South* (Palash Kamruzzaman, editor), Routledge, Taylor and Francis Group, UK, 70–87.

Blau, Francine D. and Lawrence M. Kahn (2017). The Gender Wage Gap: Extent, Trends, and Explanations. Journal of Economic Literature, 55(3): 789–865. Available online here.

Bosetti, V., Cattaneo, C. and Verdolini, E. (2015). Migration of skilled workers and innovation: A European perspective. Journal of International Economics, 96(2), 311–322.

Decent Work and the Sustainable Development Goals: A Guidebook on SDG Labour Market Indicators, Department of Statistics (2018). Geneva: ILO, 2018. Retrieved from https://www.ilo.org/wcmsp5/groups/public/---dgreports/---stat/documents/publication/wcms_647109.pdf.

Erich Fromm (1942). *The Fear of Freedom*. London: Routledge and Kegan Paul.

EUROSTAT Database. Statistic explained. Retrieved from: https://ec.europa.eu/eurostat/statistics-explained/index.php/.

Faccarello, G. (1999). *The Foundations of 'Laissez-Faire': The Economics of Pierre de Boisguilbert*. Routledge.

Filippetti, A. and Peyrache, A. (2015). Labour productivity and technology gap in European regions: A conditional frontier approach. Regional Studies, 49(4), 532–554.

Goldin, C. (2014). A grand gender convergence: The American Economic Review, 104(4), 1091–1119.

Goldin, C. and Katz, L. F. (2016). A most egalitarian profession: pharmacy and the evolution of a family-friendly occupation. Journal of Labor Economics, 34(3), 705–746.

Gorter, C., Nijkamp, P. and Poot, J. (Eds.). (2018). Crossing borders: Regional and urban perspectives on international migration. Routledge.

Graf, Nikki, Anna Brown and Eileen Patten (2019). The narrowing, but persistent, gender gap in pay, Pew Research Center, Washington, DC, https:// www.pewresearch.org/fact-tank/2019/03/22/gender-pay-gap-facts/.

Gu, W. and Yan, B. (2017). Productivity growth and international competitiveness. Review of Income and Wealth, 63, S113–S133.

Hadzhyev, K. S. (1991). The concept of civil society: ideas and main milestones of forming. Kontseptsyia hrazhdanskoho obshchestva: ydeinye ystoky y osnovnye vekhy formyrovanyia. *Voprosy filosofii* (7), 30.

Hegel, G. W. F. (1830). Hegel's Philosophy of Mind: Part Three of the Encyclopaedia of the Philosophical Sciences, trans. W. Wallace.

Hegel, G. W. F. (1990). The philosophy of Law [Fylosofyia prava]. Moscow: Mysl.

Holenkova, Z. T. (1997). Civil society [Hrazhdanskoe obshchestvo]. *Sotsyolohycheskye yssledovanyia* (3), 26.

Holzmann, R. and Munz, R. (2004). Challenges and opportunities of international migration for the EU, its member states, neighboring countries, and regions: a Policy Note (No. 30160). The World Bank.

Jones, C. I. (2016). The facts of economic growth. In Handbook of macroeconomics (Vol. 2, pp. 3–69). Elsevier.

Juan Linz, J. Linz and A. Stepan (eds.) (1978). The Breakdown of Democratic Regimes. Baltimore: Johns Hopkins University Press.

Lipton, M. (1980). Migration from rural areas of poor countries: the impact on rural productivity and income distribution. World development, 8(1), 1–24.

Locke, J. and Kelly, P. H. (1991). Locke on Money (Vol. 1, pp. 205–342). Oxford: Clarendon Press.

Lundborg, P., Plug, E. and Rasmussen, A. W. (2017). Can Women Have Children and a Career? IV Evidence from IVF Treatments. American Economic Review, 107(6), 1611–1637.

Martin, J. D., Petty, J. W. and Wallace, J. S. (2009). *Value Based Management with Corporate Social Responsibility*. Second edition, 1945: Oxford University Press on Demand.

Marx, K. (1867). Das Kapital: Kritik der politischen Ökonomie, Vol. 1. Hamburg, Germany: Otto Meisner, 40.

Marx, K. (1887). Capital: A critique of political economy (Portable Document Format (PDF) ed., Vol. 1). (F. Engels, S. Moore, and E. Aveling, Trans.) Moscow, USSR.

Neumark, D., Bank, R. J. and Van Nort, K. D. (1996). Sex discrimination in restaurant hiring: An audit study. The Quarterly Journal of Economics, 111(3), 915–941.

New York (NY): United Nations (2015). (https://sustainabledevelopment. un.org/post2015/transformingourworld, accessed 5 October 2018).

Olivetti, C. and Petrongolo, B. (2017). The economic consequences of family policies: lessons from a century of legislation in high-income countries. The Journal of Economic Perspectives, 31(1), 205–230.

Rawls, John (1971). *A Theory of Justice.* Cambridge, Mass.: Harvard University Press.

Resolution, UN (2018). *Global indicator framework for the Sustainable Development Goals and targets of the 2030 Agenda for Sustainable Development.* New York: United Nations.

Rienner, Lynne (1989). L. Diamond, J. Linz, S. M. Lipset (eds.), Democracy in Developing Countries, 4 vols. Boulder, Col.

Sachs, J., Schmidt-Traub, G., Kroll, C., Durand-Delacre, D. and Teksoz, K. (2016). An SDG Index and Dashboards – Global Report. New York: Bertelsmann Stiftung and Sustainable Development Solutions Network (SDSN).

Salamon, L. M. and Anheier, H. K. (1996). The nonprofit sector: A new global force. Working Papers of the Johns Hopkins Comparative Non-profit Sector Project, no. 21 (L. M. Salamon and H. K. Anheier, Eds.). Baltimore, MD: The Johns Hopkins Institute for Policy Studies.

Sartori, Giovanni (1987). *The Theory of Democracy Revisited.* Chatham, N. J.: Chatham House, pp. 383–93, pp. 357–62, 476–9.

Sartori, Giovanni (1991). Rethinking democracy: bad polity and bad politics, International Social Science Journal, Published quarterly – by Basil Blackwell Ltd for UNESCO, Vol. XLIII, No. 3.

Scherrer, J. (1967). *Physiologie du travail (ergonomie).* Paris: MASSON&C.

Scherrer, J. (1973). *Fyzyolohyia truda.* Moscow: Medicine.

Shybanoff, G. P. (1983). Quantitaive assessment of human activity in the systems human – technics [Kolychestvennaia otsenka deiatelnosty che-loveka v systemakh chelovek – tekhnyka]. Moscow: Mashynostroenye.

The Sustainable Development Goals Report (2018). Sustainable Development Goals Knowledge Platform. Retrieved from https://unstats.un.org/ sdgs/report/2018.

United Nations (2015). Transforming our world: the 2030 agenda for sustainable development.

United Nations (2016). Sustainable Development 17 goals, *Learn about SDGs*. Retrieved from http://17goals.org/.

Vinogradoff M. I. (1969). Gudelines on Physiology of a Labour [Rukovodstvo po fyzyolohyy truda]. Moscow: Medicine.

Volkov, N., Popov, O. and Samborskyi, A. (2003). Pulse crietrions of energetic value of an exercise [Pulsovye kryteryy enerhetycheskoi stoymosty uprazhnenyia]. *Fiziolohiia cheloveka*, 29(3), 98–103.

Volynka, K. (2003). Theory of a state and low [Teoriia derzhavy i prava] Kyiv: MAUP.

Waldfogel, J. (1998). Understanding the "family gap" in pay for women with children. The Journal of Economic Perspectives, 12(1), 137–156.

Williams, A. M. (2009). International migration, uneven regional development and polarization. European Urban and Regional Studies, 16(3), 309–322.

Williams, A. and Baláž, V. (2014). International migration and knowledge. Routledge.

Zaitsev, O. (2004). Labour value evaluation [Yschyslenye velychynû stoymosty produkta truda]. *Visnyk Sumskoho derzhavnoho universytetu. Seriia Ekonomika*, 6(65), 159–165.

Zaitsev, O. (2010). The development of the economy thoery principles [Razvytye pryntsypov эkonomycheskoi nauky] *Visnyk Sumskoho derzhavnoho universytetu. Seriia Ekonomika*, 2(1), 5–21.

Zaitsev, O. (2016). Objective value of commodities, its definition and measurement. *Scientific Letters of Academic Society of Michal Baludansky*, 4(6), 184–186.

7

The Mechanism of Venture Capital Funding of Innovative Enterprises of Ukraine Under Conditions of Globalization

Iryna D'yakonova[*], Olena Obod and Polina Rodionova

Department of International Economic Relations,
Sumy State University, Ukraine
E-mail: i.diakonova@uabs.sumdu.edu.ua
*Corresponding Author

7.1 Introduction

In the modern world economy significant changes have taken place in the capital international movement, and, in particular, new subjects have occurred – participants in the international capital market – pension funds and insurance companies, hedge funds, direct investment funds, and venture capital funds. The latter type of funds plays a special role in the development of high-tech and innovative companies that provide the development of scientific and technological revolution. Information analysis and generalization on the issues of the activity of venture capital funds in the capital market takes on great importance for activities in the international investment market.

7.1.1 The Market of Venture Capital in the General Structure of the Financial Market of Ukraine: The Functions and Development Trends

Nowadays, the task of increasing the competitiveness of Ukraine's economy is of particular importance due to not only the resource and raw material potential, but also the organization of production of high-tech original products and services based on the use of advanced technologies and highly effective methods of economic management. That is why, at the present stage, the development of practical recommendations for the improvement of the mechanism of venture capital funding of innovative enterprises under the conditions of globalization is becoming of great current importance for Ukraine.

The modern paradigm of global market economy is formed taking into account the peculiarities that are features of the current state of the Ukrainian economy. It is quite difficult for an innovative firm to function and develop in conditions of uncertainty, rather than in the conditions of developed market economy. With increasing frequency, the use of opportunities of modern innovation management brings the economic entity a huge economic effect at a relatively minimal cost. But the potential of the science and its practical implementation are quite different things. The economic conditions in Ukraine today differ from the conditions of countries in the European Union (EU), so the use of the potential of innovation management should take into account as much as possible the features, traditions and specifics of the entire country in general, and each region and the labour collective in particular.

The systematization of the proceeded literary sources shows that the scientific and technical sphere of Ukraine was not fully ready for work in the conditions of market reforms. Financing innovative activity of enterprises with the help of venture capital faces a lot of difficulties. Today, in Ukraine there is no system that facilitates the commercialization of research results, as well as their offers on the market. With intellectual and innovative potential, enterprises are trying to realize them at their own risk by exchanging them for investments. At the same time, they operate independently, without the efficient infrastructure and regulatory and legal framework necessary for such an exchange. Conditions for the development of innovation processes are shown in Figure 7.1.

The most important structures of the national systems in the EU countries are small and medium enterprises, as well as corporations of the entrepreneurial sector. They have to find the means to finance their research and at the same time, embodying scientific outputs and inventions in real products and technologies, assuming economic responsibility for the technical progress in the country. If large corporate structures can provide their scientific and research units with financial resources, then small and medium enterprises, which are created by direct researchers for the implementation of a particular innovation project, have no such resources. At the same time, small businesses are the most creative, innovative and mobile in addressing

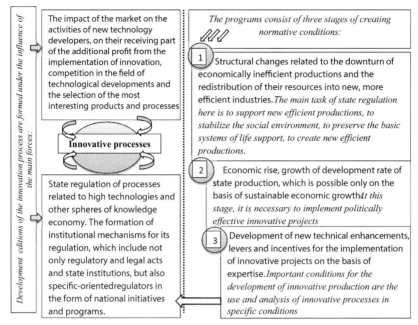

Figure 7.1 Conditions for the development of innovation processes taking into account current market trends.

innovation development issues. They respond most quickly to the needs of the innovation market and experience all the requirements and trends of innovation processes.

The main rule of the innovation process is "the development of technologies is the main driving force of economic growth". In this regard, innovation activities should take into account the following:

- A new technology appears not independently but is always associated with others;
- Each bundle consists of a number of complementary basic technologies;
- Each basic technology is the core of a number of applied technologies;
- Basic technologies is the basis of new industries;
- Applied technologies are used for modernization of the industry, therefore their use is obligatory for the entrepreneur.

Based on the main features and trends of modern innovation processes, which are the drivers of scientific and technological progress in the world and change not only the structure of production but also the structure of demand, the world has obtained a special kind of financial capital, which is aimed specifically at the field of innovative entrepreneurship, where it is possible to provide high level of return on invested capital, if the object of investment is correctly determined – an innovative company or an innovative project, which at this moment need financial resources for its formation and realization of progressive ideas. Therefore, the main source of regional and national competitiveness, economic growth of the country becomes actively support of innovative ideas. Venturer business serves as the core component of innovation development in contemporary market relations.

Intensification of innovative processes in the modern market requires substantiation of the possibilities of attracting all possible sources of funding of innovation projects and their diversification, the main objective of which is to minimize risks. Innovation should have such a degree of novelty that contrasts the associated risks with attractive returns as a result of commercialization in the future. Innovation reaches novelty recognized by the market correspondingly, when there is an understanding of its practical utility, consumer characteristics, among which new, non-traditional, more attractive properties must necessarily be of characteristics in comparison with existing analogues. But the main emphasis of the use of venture capital is on the commercial effectiveness of the innovation implementation (UAIB, 2018).

The development of the venture capital market is an integral part of the innovative development of the country's economy. The urgent tasks of our country and government include the creation of a national innovation system with the appropriate legal, organizational, and financial infrastructure. The country should actively influence the processes of innovation, taking organizational, regulatory and supervisory functions, as it is in other countries (Manayenko and Kravets, 2018).

The financial market performs extremely important functions in the market economy. It is a providing structure, primarily for the finances of business entities that are the core area of the financial system. Their financial activity begins with the formation of resources. The purpose of the financial market is to provide the enterprises with the proper conditions for attraction of necessary funds and sale of temporarily vacant resources. Thus, enterprises in financial as well as in other markets, are practically equivalent, both in the role of the buyer and the seller of the resources (Zingales and Rajan, 2004). Any market always has at least two definitions: the first as a set of relationships, the second as a set of institutions.

The investment market is a set of economic and legal relations between investors and/or investment activity participants concerning investment assets (corporate rights, financial instruments, fixed-capital assets, intangible assets). If we recall that the participants in the investment activity are those who "ensure the implementation of investments as account executives or in execution of the investor's instructions," the appropriate investment institutions can be found among them. Then the investment market can also be defined as a network of investment institutions that implement contractual relationships under investment asset management by means of the provision of investment services. These investment institutions include investment companies; collective investment scheme; venture funds; direct investment funds.

Considering venture capital in the system of innovation processes in the economy, it is important to mention that it is not only closely related to innovation, but it is also a critical factor in the innovation process.

The point of view of the scientists who consider venture capital as a special functional form of financial capital deserves attention. After all, the risk character of venture capital makes it possible to distinguish it as a special kind of capital. Venture capital is an investment resource for providing innovation activities; reflecting the economic relations between the subjects of venture business; is naturally of risky character.

Venture capital, by definition of the European Venture Capital Association (EVCA), is a share capital provided by professional financial companies that is invested in start-up innovative enterprises that are developing and exhibiting a potential for significant growth (UAIB, 2018). The fundamental difference of venture capital from traditional investment is that it is primarily invested in an idea, a project with an increased risk. Venture capital addressees are venture capital firms that are not required either to pay interest or return the received amounts. Therefore, the investor's interest is satisfied by the acquisition of rights to all innovations, both patented and unpatented know-how, and profit from the implementation of scientific and technical developments. Funds for venture capital funding are accumulated by special financial and banking institutions.

At the present stage, fundamental scientific research in the whole world is carried out at the expense of public funds and funds of large corporations. The latter are most interested in innovations, while small innovative companies develop them, which respond to the needs of the market more quickly, but they need special funding sources – more affordable in cost and time (long-term basis). The venture capital is exactly this source of funding, which is invested in small and medium-sized innovation firms, therefore contributing to the transformation of R & D results into commercialized products and services in the market.

Since both venture capital and portfolio of venture capital investors are formed based on different sources and investment objects, various financially powerful companies are involved as investors, depending on the importance of the problems for the global or national economy, being solved when forming certain volumes of investments for financing of projects. So, if problems important for humanity are solved, which do not bring venture capital investors the expected profits (ensuring environmental cleanliness, the prevention of existing threats to mankind – floods, typhoons, eruptions, epidemics, etc.), then powerful investment institutions are involved as investors, which have possibility to finance such projects on the basis of public and private partnership or using new organizational financing mechanisms.

In order to effectively manage the development of the venture capital market, its participants should be further substantiated, in particular, venture capital market subjects and objects in the context of their functions and tasks that are being solved. The object is the process of functioning of the mechanism of venture capital funding of innovative enterprises of Ukraine under conditions of globalization. The subjects of the venture capital market are:

- Venture capital investors – legal entities and individuals that provide financial resources for risk investments in the company;
- Venture capital companies – those that manage the money received from investors and place them in various innovative projects;
- Recipients of risk capital – companies funded by venture capital.

Technological revolutions, which led to the transformation of entire economy areas, were headed by firms based on venture capital. For example, firms that went ahead of each new generation of computer technologies (personal computers, software, etc.) were funded by venture capital. The venture capital market is divided into formal (institutional) and informal sectors. The institutional sector is mainly represented by venture capital funds, and the informal – by individual investors. The essence and objectives of the venture capital market are shown in Figure 7.2. Besides, the venture capital market, which belongs to the type of imperfect competition markets, is typically characterized by asymmetry: uneven distribution of information between the subjects – venture capital investors and venture capital consumers (recipients). As a rule, recipients of venture capital get the information edge. In this market environment, supply and demand are balanced by market intermediaries – venture capital companies.

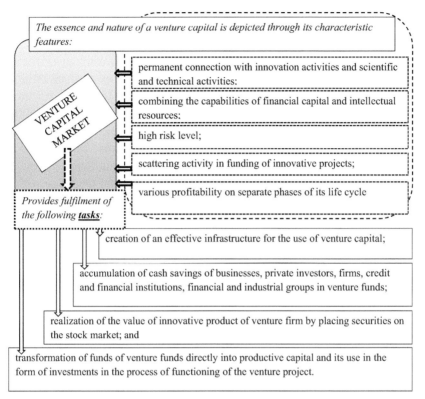

Figure 7.2 The essence and objectives which are ensured by the effective activity of the venture capital market.

Consequently, the efficient development of the venture capital market in both the EU and Ukraine is in need for the formation of its own hierarchy and a network of national elements in the configuration of the national innovation system, which reflects its national peculiarities, in particular, the combination of public and private investments in innovation activity funding, the ratio of large and small companies in the innovation process, the ratio of fundamental and applied research and development, the dynamics of development and the sectorial structure of innovations activities.

The system of venture investment is a universal mechanism for funding the creation of an innovative product on the basis of the integration of intellectual labour and capital, an effective form of ensuring the implementation of a global strategy of social and economic progress, based on scientific achievements and placing of their results on the consumer market.

7.1.2 The Institutional Foundations of the Investment Funding Mechanism of Venture Business

The market for venture investment is part of the world investment market, which differs from the latter by degree of risks in the process of investing by its very nature. This means riskier areas of industry and start-ups for the investments, which also with a high degree of probability are not successful in most cases. At the present stage, the introduction of an innovation-oriented system of economic development for Ukraine has a number of problems. The main problem is the problematic financial support of innovative projects of high risk. Unlike other spheres of economic activity, the degree of risk in innovation is super-high, which makes it impossible to participate in traditional loan capital in financing risky innovation projects. For effective results of financial support for innovation processes, new alternative sources of investment resources are needed, among which it is first of all appropriate to consider the development of venture capital funding, which is one of the important areas of support for medium and small innovative entrepreneurship, research funding, etc. in developed countries.

In Ukraine, the market for venture capital is presented in two types: informal (in the form of individual investors) and formal (represented by venture funds). The structure of the Ukrainian venture capital market is shown in Figure 7.3. Formal and informal sectors play complementary relationship roles. Investments in the informal sector are particularly important at the initial stages of the development of start-up firms when they need seed capital to develop the concept of a product and a prototype, while the formal sector is at a stage of a rapid growth of the company, when funds are needed to expand production and increase sales.

According to the association of business angels U-Angel, the average check of the Ukrainian business angel in the transaction is 20 to 60 thousand USD. Ukrainian angel investors are the most interested in e-commerce, online services, software, large data, fintech and analytics. For comparison: in a closed club for private investors iClub, which was created by TA Ventures, the minimum investment in an agreement starts at 25 thousand USD.

At the beginning of 2017 Ukraine had no more than 50 system business angels. In the same period, the total of about 260 thousand dollars was invested in Europe. U-Angel state that there is a growing number of syndicate transactions in Ukraine (Novak, 2018). Despite the negative economic factors affecting the investment climate in Ukraine, today it is named one of the most attractive countries for investors: As reported UNIAN, the Cabinet of

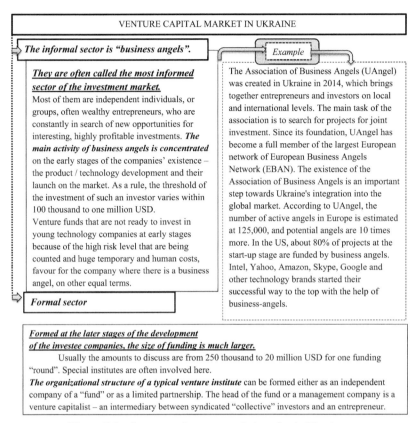

Figure 7.3 Structure of venture capital market in Ukraine.

Ministers of Ukraine expects that inflow of foreign direct investment by the results of this year will amount to 4.5 billion USD, in comparison with 3.8 billion USD engaged in 2016. The government forecasts economic growth of 1.8% in 2017, 3% in 2018, 3.6% in 2019, and 4% in 2020 (Unian, 2017a). The index of investment attractiveness of Ukraine according to the version of the European Business Association dates back to 2008. The lowest indexes were recorded in 2014 and early 2015, which was largely due to political instability, economic downturn and the beginning of active military actions in the east of Ukraine. Analysis of the data shown in Figure 7.4 demonstrates that the index is 0.28% in 2016, and it is 3.15% in 2017.

Company senior executive teams point out that real business deregulation has begun in Ukraine, electronic government services have been introduced,

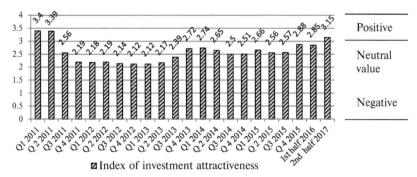

Index of investment attractiveness

Figure 7.4 Index of investment attractiveness of Ukraine for the period from 2011 to 2017 (Unian, 2017b).

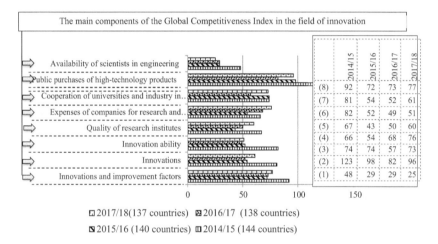

☐ 2017/18(137 countries) ⊠ 2016/17 (138 countries)

◨ 2015/16 (140 countries) ▥ 2014/15 (144 countries)

Figure 7.5 Global Competitiveness Index of Ukraine in Innovation (Derzhavna sluzhba statystyky Ukrainy, 2016).

construction permits have been simplified, and the institution of private court judgement agents has been introduced.

According to the results of the analysis of the position of Ukraine on the Global Competitiveness Index, which are shown in Figure 7.5, it can be seen that the national scientific and technical base and the available number of scientists and engineers occupy high positions in comparison with other indicators. The quality indicators of research institutes are also high, which confirms the potential and good level of ability to innovate.

However, indicators of the state of cluster development and the decline in the quantity of state purchases of high-tech products are the lowest, which

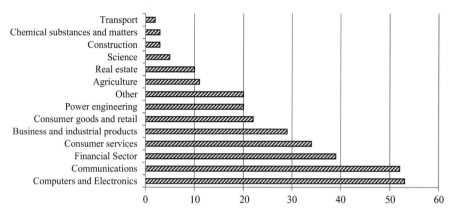

Figure 7.6 Advantages of the main players of the market of direct venture investments of Ukraine in 2016.

indicates the lack of a coherent mechanism for cooperation between research institutes and industrial enterprises and the country's lack of interest in developing the innovative environment on the market in recent years, which slows down the diffusion of innovation processes and their implementation.

The structure of the Ukrainian economy is focused on raw materials, in particular on the mining industry, agricultural industry and export of agricultural products, while the economy of the European Union is characterized by a reorientation from high-tech material production to the development of the social sphere and the sphere of services (Manayenko and Kravets, 2018). The data confirmed by the expert company UVCA is shown in Figure 7.6.

World experience has shown that venture capital in CEE countries is more stable than conventional investments, since it focuses mainly on high technology and more in the fields of computerization and electronics, especially when it comes to the virtual field.

Spheres of energy and chemical industry, which have a long investment cycle from the point of view of venture capital, remain vulnerable in terms of venture capital investment. This is a rather unfavourable circumstance for any state, therefore, state regulation should be involved here to a greater extent, but these sectors are more important in the long run for the growth of the economy as a whole. The data shown in Figure 7.7 illustrate the total number of venture funds in Europe from the first quarter of 2015 to the fourth quarter of 2017.

The number of venture capital funds in Europe fluctuated throughout the analysed period reaching a total of 26 in the fourth quarter of 2017. The

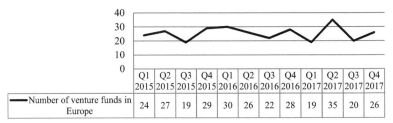

	Q1 2015	Q2 2015	Q3 2015	Q4 2015	Q1 2016	Q2 2016	Q3 2016	Q4 2016	Q1 2017	Q2 2017	Q3 2017	Q4 2017
Number of venture funds in Europe	24	27	19	29	30	26	22	28	19	35	20	26

Figure 7.7 The number of venture funds in Europe for the period from first quarter of 2015 to the fourth quarter of 2017 (Statista Database, 2018).

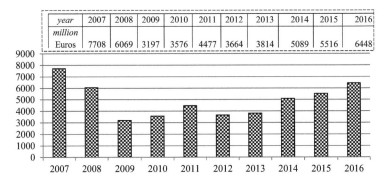

year	2007	2008	2009	2010	2011	2012	2013	2014	2015	2016
million Euros	7708	6069	3197	3576	4477	3664	3814	5089	5516	6448

☒ Total value of venture capital funds attracted in Europe from 2007 to 2016

Figure 7.8 The total value of venture capital funds attracted in Europe from 2007 to 2016 (Lapko, 2006).

Source: https://www.statista.com/statistics/421411/number-of-venture-capital-funds-europe/.

largest number of venture funds in Europe was detected in the second quarter of 2017, when 35 venture funds were registered. The venture capital fund is part of an investment fund that manages the monetary funds of investors, which are predominantly represented in small and medium-sized enterprises (SMEs), as well as in start-ups. It is characterized by high growth potential and higher risks for return. The total value of investments in private capital of venture capital in Europe from 2007 to 2016 is described in Figure 7.8.

The data shows that the total value of venture capital investments fluctuated during the observation period, reaching almost 6.45 billion Euros in 2016. The largest value of investments was recorded in 2007, when venture investments amounted to over 7.7 billion Euros. The most active investors for European venture capital companies in 2018 are: High Venture Capital, Bpifrance Investment SAS, Funding London, Local globe LLP, etc., as shown in Figure 7.9.

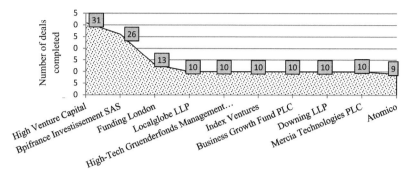

□ Leading investors for European venture capital companies

Figure 7.9 Number of completed deals of leading investors in the venture capital market of Europe as of the second quarter of 2018.

In the second quarter of 2018, High Venture Capital is the most active investor for European venture capital companies (by number of deals). The leading investor in the venture capital market completed 31 deals under contract. Localglobe LLP, High-Tech Gruenderfonds, Management GmbH, Index Ventures, Business Growth Fund PLC, Downing LLP, Mercia Technologies PLC completed 10 deals under contract in the second quarter of 2018.

The largest venture funds in the world are investing in large-scale industrial projects that are more closely linked to high technology. Unlike world practice, venture investments in Ukraine go to the IT industry, and not to the technical component, but to the virtual one. Therefore, this is an evidence of low venture diversification in Ukraine on the one hand, and the lack of desire of international investors to get engaged in any area other than IT on the other. The most active participants in the venture market, which invested in Ukrainian projects are Almaz Capital, AVentures Capital, BeValue, Digital Future, Chernovetskyi Investment Group, Empire State Capital Partners, GrowthUP+ VC Fund, Imperious Group, Internet Invest Group, Horizon Capital, KM Core/Borsch Ventures, Runa Capital, TA Venture, TMT Investments, SMRK VC Fund, and others.

According to a poll of key players in the market for direct and venture investments by Ernst and Young experts in April 2016, investors continue to focus mainly on stable large markets where business can be scaled up. Geographical preferences of the main players in the market for direct and venture investments of Ukraine are shown in Figure 7.10. Respondents showed the greatest interest in the markets of developed countries, namely

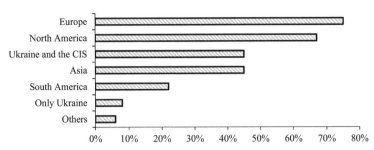

Figure 7.10 Geographical preferences of the main players in the market for direct and venture investments of Ukraine.

75% would like to invest in Europe, 67% – in North America. Despite this, the CIS countries' as well as Asia's interest in Ukraine remains at a rather high level (46%).

Only 8% of the respondents are ready to focus only on the domestic economy. Since work on projects invested by respondents is conducted simultaneously in different countries, the survey focused on the markets where final projects are being implemented. The respondent investors also shared their views on the state of the projects under consideration in the Ukrainian market. According to 87% of respondents, their number is insufficient, which means significant potential for market growth. In addition, 43% of respondents believe that the quality and preparation of projects in Ukraine are good or at least satisfactory. Despite these negative features, Ukraine has good reason to hope for success. First of all, due to the industries with potential growth and a large number of promising projects that needs funding and is potentially beneficial to a venture investor. Therefore, an urgent need is to develop a competent government policy to stimulate venture business taking into account the leading foreign experience.

7.1.3 Improvement of the National Mechanism of Venture Capital Funding of Innovative Enterprises of Ukraine Under Conditions of Globalization

One of the most important conditions for the existence and development of an innovative economy is the maturity and stability of the institutional environment, which ensures the interaction of business entities in the conditions of intense technological changes. At the same time, the old institutions begin to intensively modernize under the influence of new economic goals, driven by technological changes. There emerge mechanisms that make it possible to continuously develop institutions that stabilize and regulate the economic environment of the new economy. The most important institutions of the innovation economy include such as R & D institutes, institutes of intellectual capital production, funding institutions for fundamental research, informal institutions that determine the creation of innovations, in particular, culture, mentality, reaction to the action of the external environment.

Unlike classic venture funds, Ukrainian venture capital does not focus on innovation projects. It is more attracted by the implementation of medium-risk investment projects using financial assets and real estate operations (Lapko, 2006). The efficient functioning of the venture capital industry is able to provide the most dynamically developing innovative enterprises with investment, thus contributing to the competitiveness of the national economy (Zinchenko et al., 2004).

It is impossible to develop a competitive domestic industry without state support, the development of innovations and tools for the development of innovation. Abroad, in the USA and Europe, various mechanisms for funding business projects are actively used to introduce new technologies into production. For example, the world's leaders in the computer industry Microsoft, Apple, Intel took their current position largely because of venture capital investments in the early stages of their development. Venture business is an entrepreneurial activity associated with the implementation of risky projects, risky investments mainly in the field of scientific and technical innovations (Zinchenko et al., 2004).

Foremost, the general economic situation in the country and not the presence of venture ideas, influences venture investors (Kirsanova et al., 2011). In Europe, unlike the USA, there is no division into venture funds and direct investment funds. Also, the big difference from the American venture business is that European venture funds invest in virtually all sectors of the economy. In recent years, the reorientation in European venture capital investments moves to the technological sector (Kirsanova et al., 2011).

In the European countries the state is the driving force behind venture business and a stimulus factor, unlike private capital, which prevails in the countries of North America. In the American model of venture financing, the main purpose of the state is to integrate and promote the interaction of the industrial and scientific sectors. The state is assigned the role of "night watchman", which does not significantly interfere with the promotion of venture projects and performs legislative regulation of venture processes. The development of a mechanism of state support for venture business in Ukraine is extremely important, as with the country's economy becoming increasingly sensitive to global economic processes, venture capital business is at an early stage, and it may remain there if it does not use foreign experience of countries in stimulating venture capital investments.

An integral part of the venture capital investment and development of the venture industry in Ukraine is the state regulation of venture activity. The reason is that the priority directions of government policy in most developed countries are innovation, because they are the basis of national independence, security and economic uplift. The lack of statistical information on the sectors receiving venture capital investments emphasises the inappropriate use of the organizational form of venture funds: optimization of asset management of financial and industrial holdings and reduction of tax take. It is the state that should promote the creation of all the necessary conditions for stimulating innovation activity by means of the formation and implementation of generally accepted rules, principles and mechanisms, which will promote the development of the innovation industry. The reasons for state regulation of venture investments are shown in Figure 7.11. Therefore, the most important condition for the evolution of the effectiveness mechanism in the institute of current venture capital funding is the creation of a special regulatory and legal framework that ensures a combination of national priorities in the field of scientific and technological development and the development of knowledge-based industries.

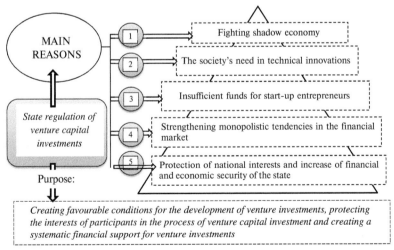

Figure 7.11　The main reasons for state regulation of investments in modern conditions (Polishchuk, 2017).

It is greatly necessary to arrange the institutional integration of small knowledge-based venture business into the system of national knowledge-based industry using special state programs as the main development tool. The main ways to improve the institutional environment of venture business are described in Figure 7.12.

Nowadays, the process of implementation of the venture capital funding institution is currently unsystematic in Ukraine. Separate legislative acts are introduced that contradict other laws and branches of law. In order to accelerate the process of implementation and development of institutions that

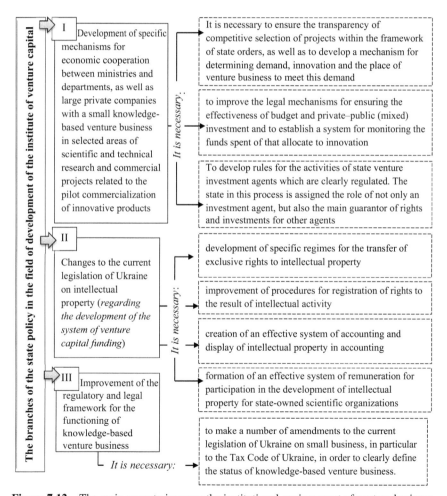

Figure 7.12 The main ways to improve the institutional environment of venture business.

coordinate all elements and links of the funding mechanism of domestic venture business, a clear state science and technology policy is needed in combination with balanced market-oriented corporate legislation.

The achievement of the set goals will be possible under the condition of successive implantation of certain institutes and procedural mechanisms of interaction. Thus, objectively existing legal problems hindering the development of the venture industry can be solved by adopting batch changes to existing regulations. In general, there is no need to adopt a special law on venture capital activities. Venture capital activities as a type of business do not require any special regulation. The introduction of a special law, declared as a contribution to the development of venture capital activity, can lead to a reverse effect.

The development of a mechanism of state support for venture business in Ukraine is extremely important, as with the country's economy becoming increasingly sensitive to global economic processes, venture capital business is at an early stage, and it may remain there if it does not use foreign experience of countries in stimulating venture capital investments. Before developing a mechanism for state support, one must understand towards which model Ukraine should be oriented: American, European or mixed.

The American model is distinguished by its liberalization and minimal state interference in the regulation of venture capital investment. Major investments are made by private investors, corporations, universities, etc. In Europe, the regulatory model is based on much more state intervention: the development of national and pan-European venture investment programs, which are more oriented towards bank capital. In turn, the creation of centralized venture funds is common in Europe such as "funds of funds", to accumulate resources for their further redistribution among diverse venture funds. However, with all the benefits of the American model of venture regulation, Ukraine will not immediately be able to copy it, because an appropriate culture of doing business has developed from the very beginning of the country's existence in the United States, while Ukraine still cannot formulate its own principles of its conducting and come to at least general features of corporate governance that would characterize Ukrainian entrepreneurship.

That is why the most effective is the establishment of a mixed system of regulation of venture business in Ukraine. This process should take place through the primary orientation of the European model with the corresponding creation of a large number of state regulatory authorities. After venture investments in Ukraine begin to move to a qualitatively new level, state regulation should be weakened, turning to corporate forms of investing in

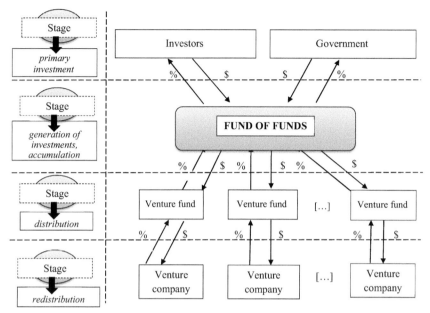

Figure 7.13 Scheme of venture capital funding of funds.
Source: Compiled by the author.

Ukrainian venture business. Accordingly, one of the first steps should be the creation of a fund of venture capital funds. These institutions do not necessarily have to be public, they can also be exclusively private or with a mixed ownership structure. As for public funds, they can become an important initial source of the capital that could contribute to the accumulation of investment experience and its gradual transfer to the private sector. In order to understand the advantages and disadvantages of funds in the financing process, we give a schematic diagram of this process shown in Figure 7.13.

The benefits of the funds are that they are the accumulators of significant capital, which are further effectively distributed and redistributed between venture companies. Their efficiency is particularly noticeable in underdeveloped investment markets, where small investors who do not have significant capital participate in the stage of primary investment. Instead, the intermediate stages (generation and distribution) lose their effectiveness when the corporations interested in investing emerge during the initial investment stage. Therefore, direct investment from the 1st to the 4th stage in this case becomes cheaper without losing its efficiency. Therefore, it is impossible to miss the 2nd and 3rd stages of financing on the non-perfect investment market in Ukraine; however, with the development of the investment market, these stages will have to be dropped, as they will reduce the investment attractiveness of the market as a whole due to lower investor's equity.

The next step is to create a special venture investment committee (department), which will deal not only with the financing of funds and venture funds at the state level, but also will engage in marketing of the national market abroad for more active involvement of foreign investors. An important feature of both European and American markets is the activity of insurance companies and pension funds on the market of venture investments. This sphere in Ukraine is extremely undeveloped, and insurance companies with functions of pension funds do not exist at all. Therefore, the problem of reforming the pension system is extremely acute, because, in addition, there is a permanent deficit of the pension fund in the country. The mechanism of funding of innovative enterprises of Ukraine is described in Figure 7.14.

Regarding the method and details of improving the financing mechanism of venture business, the general rules for building a state policy in the field of creating a venture financing institution in the country are also essential:

1. The policy objectives related to the innovation financing should be realistic; they should take into account the prevailing background conditions for the four stages of the financing cycle of innovation developments and consider the link between the various factors of supply and demand.
2. The policy objectives should clearly identify the types of enterprises that should receive support.
3. The success of the study of the experience in other countries, the implementation of programs successfully carried out in these countries, depends on whether investors clearly understand what actions were taken in the past.

In conclusion, it can be noted that the reform of the venture market in Ukraine is complex and strongly connected to the stability and growth of all spheres of the Ukrainian economy as a whole. The key to the success of a venture business is the versatility of funding sources. In Ukraine, in fact, investments are mainly made by private investors or companies; therefore, the activation of alternative sources of financing is required. In the long run, the transition to a "free venture market" should be ensured alike the American model. The free market will allow the most efficient allocation of resources to minimize the bureaucratic system.

Ukraine has a significant human potential in the intellectual dimension, which has not yet been fully realized. Moreover, its drain is constantly going

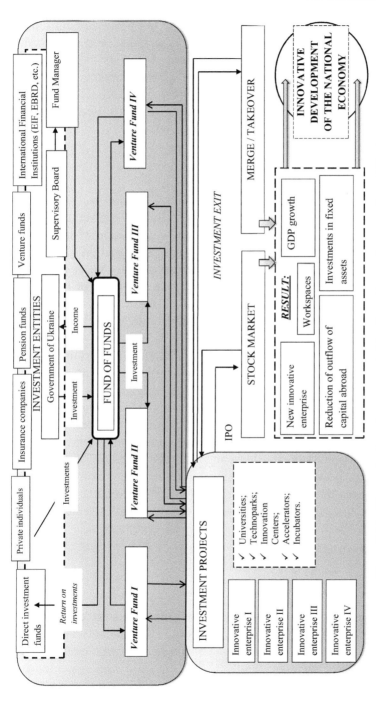

Figure 7.14 The mechanism of venture capital funding of innovative enterprises of Ukraine.

Source: Compiled by the author.

beyond Ukraine. That is why the realization of the project of reforming the of venture investments market in Ukraine is strategic, and it forms the future competitiveness of the Ukrainian economy for years to come. Until the Ukrainian elites recognize the fact that the global economic system has already entered the phase of the latest high technologies, Ukraine's economy will continue to remain in periphery, constantly losing its scarce advantages that it still has.

7.2 Conclusions

Nowadays, the development of venture financing in Ukraine is characterized by the need for innovative development, which is also aimed at venture capital funds. Venture capital positively influences the acceleration of innovation processes in the economy, when it performs its main function – investing in small innovative enterprises that realize their progressive ideas, providing positive innovations in the structure of production in favour of the development of high-tech structures. The essence of the influence of venture capital on innovation processes is manifested only through its main function – venture capital investment.

Based on the analysis of theoretical and analytical data in the area of innovation activity, we can come to the conclusions that for the accumulation of the necessary amount of venture capital in Ukraine it is necessary to reform the national economy: to stimulate entrepreneurial innovation activity; to form a complete innovative infrastructure; to develop stock and insurance markets, and banking system; to provide proper conditions for the effective operation of private investors (business angels, venture funds). The regulatory framework should provide government support for the venture capital system, as in the EU, when the state plays a driving force in the venture sector. This will create conditions for the development of various forms of venture capital investment and the targeting of venture capital investments in high-tech production.

Acknowledgement

We would like to give our special acknowledgements to our colleges at the department and beyond the University, PhD students and young scholar who took part in brainstorming sessions and motivate us for further research.

References

Anypyn, V. M. and Fylyn, S. A. (2003). The Investment and Innovations Management in small business. Moscow: «Ankyl», 360 p.

Derzhavna sluzhba statystyky Ukrainy [Governmental statistical service of Ukraine] (2016). Available at: http://www.ukrstat.gov.ua.

Drachuk, Y., Stalinska, O., and Trushkina, N. (2017). The mechanisms of stimulation of venture investments towards innovation development in the industries: the generalization of foreign experience [Mekhanizmy stymuliuvannia venchurnoho investuvannia innovatsiinoho rozvytku promyslovosti: uzahalnennia zarubizhnoho dosvidu]. *Economic discourse*, 2, 117–128.

Kirsanova V. V., Suhareva, T. O., and Kovalova, O. M. (2011). The problems of accounting of innovation activity [Problemy obliku innovatsiinoi diialnosti]. The Visnyk of Social and Economic Research, 41(2), 216–221.

Lapko, O. O. (2006). The Venture Capital as a source of financing the innovational development of economy [Venchurnyi kapital yak dzherelo finansuvannia innovatsiinoho rozvytku ekonomiky], *Economics and forecasting*, 3, 26–32.

Manayenko, I. M. and Kravets, A. I. (2018). The financing the innovation activity of the enterpresies: Ukrainian reality and the experience of EU. *Economics and management of the Enterprises* [Ekonomika ta upravlinnia pidpryiemstvamy], 15, 79–82.

Novak, N. (2018, August 17). The phenomenon of private investments [Fenomen pryvatnoho investuvannia]. *Finance.Ua*. Available at: https://news.finance.ua/ua/news/-/432627/fenomen-pryvatnogo-investuvannya-abo-yak-distaty-groshi-z-pid-matratsiv-ukrayintsiv

Polishchuk, O. (2017). The current situation of the state regulation of venture industry in Ukraine and the pathways for the further development [Suchasnyi stan ta shliakhy rozvytku systemy derzhavnoho rehuliuvannia venchurnoi industrii Ukrainy]. *Economics and organization of the management*, 3(27), 79–86.

Statista Database (2018). Venture Capital Funds in Europe in 2015–2017 period. Available at: https://www.statista.com/statistics/763156/number-of-investments-in-start-ups-in-europe/

The BVCA Private equity and Venture Capital Report on Investment Activity (2012). Available at: https://www.bvca.co.uk › documents › RIA_2012

The official site of the Analitical Center CEDOS (2018). Available at: https://www.cedos.org.ua.

UAIB, Official site of Ukrainian Association of Investment Business, (2018). Available at: http://www.uaib.com.ua/analituaib.

Unian (2017a, September 5). Ukraine was named one of the most attractive countries for investors. Available at: https://www.unian.ua/economics/finance/2117056-ukrajinu-nazvali-odnieyu-z-naybilsh-privablivih-dlya-investoriv-krajin.html.

Unian (2017b, July 3). Research data. Ukraine is the most attractive for investment in the last 6 years – the research. Available at: https://espreso.tv/news/2017/07/03/ukrayina_sogodni_naybilsh_pryvablyva_dlya_investyciy_za_ostanni_6_rokiv_doslidzhennya

Zinchenko, O. P., Ilchuk, V. P., Radziievska, L. F., and Yevtushenko, V. M. (2004). The state of the development of venture start-up organizational forms and the venture infrastructure in Ukraine and in global [Stan rozvytku orhanizatsiinykh form venchurnoho pidpryiemnytstva i yoho infrastruktury v krainakh svitu ta v Ukraini]. Kyiv: NDISEP, 80 p.

Zingales, L. and Rajan, R. (2004). Saving capitalist from the capitalists: Unleashing the Power of Financial Markets to Create Wealth and Spread Opportunity. Translated into Russian. Moscow: TEIS, 492 p.

8

Economic Restructuring of Ukraine National Economy on the Base of EU Experience

Olena Shkarupa[*], Oleksandra Karintseva and Mykola Kharchenko

Department of Economics, Entrepreneurship and Business Administration,
Sumy State University, Ukraine
E-mail: elenashkarupa@econ.sumdu.edu.ua
[*]Corresponding Author

It will consider the peculiarities of scientific and methodological approach to the modernization of economic structure of Ukraine National Economy in the context of SD goals. The analyses providing problems and prospects of restructuring of national economy according to environmental conditions as well as the role of innovations in restructuring process will be done. It will present the experience of implementation of the innovations for the SD on the base of EU experience. It also will focus on the problems of costs of economic restructuring of national economy. The research investigates the methodological basis of economic structure, which is two groups of structuring principles: (1) characterizing the dynamics of economic structure; (2) describing the statics of economic structure; the basic factors determining the economic structure are also established and conditions necessary for the formation of the optimal structure of the national economy are identified. The study of approaches to the typology of economic structure is undertaken and the most appropriate one is identified. A new type of structure – "eco-destructive", which characterizes the influence of environmental losses from the activity of economic entities on the economy and sustainable development of the country as a whole, has been suggested. An insufficient attention is paid to indicators that characterize the destructive impact of economic agents' activities on the environment, which eventually influences the estimated component of the potential in the opposite direction.

8.1 Introduction

In recent years, the concept of "green" economic growth has created new problems for scientists who are involved in the formation of the necessary prerequisites for sustainable development. The problems that arise the questions of funding, demographic and social changes, political instability, the low quality of institutions and the instability of markets, encourage scientists to debate how the state can help create an enabling environment for achieving sustainable development. The expected progressive changes in the economies of Ukraine are the cause for the need to create the prerequisites for sustainable development, which in turn envisages a transition from an orientation towards the consumption of natural resources, as materials and energy, to an orientation towards a more rational use of these with the activation of the use of environmental and social functions of nature. Of great importance in this context is the ecologically oriented development of the basic sectors of economy: organic farming, forestry, recreational complexes, tourism sector, creative economy and other sectors of the economy that form the basis of the soybean economy. These spheres of economic activity have a significant potential in increasing the competitiveness of economic systems in Ukraine.

To achieve this, it is necessary to solve the problems of assessing the process of sustainable development at the national level. Within the last few years in the Ukrainian economy significant transformation changes in the direction of "green" growth have been observed. On the background of a significant reduction in consumption and import of natural gas, during the first half of 2018, almost 270 MW of capacities was installed in Ukraine, which generates electricity from renewable energy sources at a feed-in tariff. This is 2 times more than in the first half of 2017 (127 MW) and exceeds the capacity set for the whole 2017 (257 MW). In total, from the beginning of 2015 by the end of the second quarter of 2018, 677 MW of new capacities of renewable energy were commissioned, with more than 650 million euro invested. Specific tasks and priorities of "green" economic growth in Ukraine are identified in the National Renewable Energy Action Plan till 2020, Energy Strategy of Ukraine till 2030, Energy Community Directives, EU-Ukraine Association Agreement and other strategic documents. Also, the author (Melnyk, 2016) notes that a significant number of publications on the "green" economy affect its external attributes. On his opinion, the internal mechanisms of the formation of economic systems in transition to a new state of the society remain less explored, which determines the urgency of this research.

Current approaches of formalizing the model of "green" economy in Ukraine are based on directing investment resources to the development of "key" sectors of the economy, which should provide its gradual "green" growth, namely agriculture, forestry, fishery, energetics, building, transportation, IT technology and tourism. Some approaches are based on the macroeconomical division of the economy or on the experience of other countries (Potapenko, 2012; Prushkivska, 2013, 186; Romanenko, 2016, 118; Bublyk, 2016; Gura, 2017). On the background of the lack of conceptual clarity concerning development of "green" economic model in Ukraine it is necessary to admit that the processes of transformational changes in the economy aimed at forming an effective environment for "green" growth are not completely studied yet.

8.2 Theoretical Bases of Ukrainian Economic Structure

In modern scientific literature the structure of the national economy is considered in the majority of cases from the position of the ratio between the branches of production and in the context of reproductive processes. Therefore, the optimal structure of the national economy is formed on the basis of the concept of sustainable development. This requires a revision of the approaches to typology of economic structure and considering the life cycle of each of type of economic activity.

National economy is a complex system because it combines a set of elements that function as a unit, and they are characterized by close interrelationship and coherence with the aim of promoting the realization of a common goal (Maliuk, 2008), and is defined by the parameters presented in Figure 8.1. Accordingly, analyzing Figure 8.1, it can be argued that the national economy is a dynamic category, which is formed for a long period of time under the influence of many factors, both internal (further, the features of the economy form national identities) and external (the economy corresponds to the world requirements for development and activities, the general principles of functioning, the models of leading economic systems).

The retrospective of the formation of the national economy (the degree of development of the productive forces, the propensity to innovate, the state policy, sensitivity to external shocks, etc.) plays an important role in its further stability. Along with this, in our opinion, the economic structure is the fundamental criterion that forms the basis of all further researches, developments, steps of the state towards intensive economic development of the country (for example, in the form of reforms) as a whole. The economic

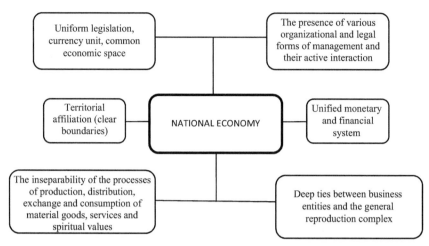

Figure 8.1   General features of the national economy (Reshetylo, 2009).

structure is the cornerstone on which the national identities of the economy are formed. There is no doubt that the national economy is a clearly structured system.

The work of economic statistician C. Clark titled "Conditions of economic progress" is evolutionary in the context of the research questions of economic structure. It proposes to structure the economy into three large sectors (the first sector is agriculture (production of raw materials), the second is industry (production of finished products) and the third – services), each of which included the industry, based on technological and economic characteristics of its development. According to the scientist, with the development of the economy and its transition to the top lines of functioning, the dominant sectors of economy are changed. Analyzing the modern structure of the world economy and the transition of countries to post-industrial development, we can confidently assert the correctness and fairness of the conclusions of Clark and his follower J. Furaste (Fourastié, 1951), who noted the prevalence of employment in the service sector. The continuation and development of the Clarke model of economic structure was also carried out by the American sociologist (Bell, 1999). The modification of his point of view on economic structure has the following form (Prushkivs'ka, 2013):

- The first sector: traditional production (raw materials presented with agriculture, forestry and extractive industry);
- The second sector: manufacturing (raw materials processing and construction);

- The third sector: production of material goods (transport, communications and utilities);
- The fourth sector: other services (trade, finance, real estate operations);
- The fifth sector: public services (health, education, research).

It is worth noting that to date the sectors modification in economic structure of the continues.

In our opinion, the economic structure characterizes the consistent development of the economy as a system and its individual components. The complexity and multidimensionality of the category under study, among other things, is explained by the presence of a large number of interrelated processes (reproductive, resource, technological, labor, investment, innovation, financial, organizational, etc.). Proceeding from this, it is proposed to consider the economic structure as an aggregate of various elements of economic system, characterized by corresponding interrelations between them and interdependence of each other and, as a result, form the basis for the system stability, stable development of the entire economy and sustainable development of the country as a whole.

The analysis of conceptual approaches and works devoted to the study of theoretical basis of economic structure makes it possible to form a list of fundamental principles that ensure the observance of features of systemic nature in further research (Table 8.1). It is worth noting the advisability of dividing all principles into two large groups: the first group comprises the principles that characterize the dynamism of economic structure, and the second group includes the principles that determine the static structure of the economy. This division is due, firstly, to the fact that the economic structure is defined by constant fluctuations that occur as a result of structural changes, and secondly, by the fact that the analysis of economic structure is always conducted for a specific (fixed) period of time and corresponding dynamic changes in the structure can be track only on the basis of its static states.

The process of forming the economic structure can be traced through structural changes that reflect the transformation of relationships between various components of economic system in space and time. Accordingly, structural shifts arise as a result of gradual accumulation of structural changes in the economy. For example, the gradual economic development of the country causes changes in the structure of production and consumption, reflected in a reduction in the share of industries that produce primary resources and an increase in the share of industries producing services. So, on the one hand, the share of primary resources in the gross domestic product is indicative of the level of technological development of the economy and its individual

Table 8.1 Principles of the study of economic structure

Group	Principle	Definition
I	Evolutionism	Characterizes the relationship between the development of internal and external environment of economic system and structural changes with the aim of preserving its integrity and the unity of its elements
	Cyclicity	Defines the economic structure as a relatively independent characteristic of the system, which in its development passes through the stages from progressive to regressive influence on the system integrity
	Polycentricity	Determines the direction of development of economic structure, proceeding from the continuity of reproduction process
II	Subjectivity	The principle that allows us to classify the economic structure depending on its component's composition, the influence of the elements on each other and on the ties quality
	Polystructural Structure	The principle that determines the complexity of economic system, its fullness with many structural elements that have a certain degree of self-regulation based on the interests and multifaceted relations of subordination in subsystems of one level to another

sectors (the smaller the share of consumption of primary resources, the more productive is the economy of the country). On the other hand, the growth in the share of services in GDP indicates an increase in the wealth of the nation and the quality of life of population, since a poor society requires a much smaller list of services. So, we can conclude that it is the wealth of a nation and the degree of economic development that determines the economic structure and not vice versa.

In professional scientific literature one can find a considerable number of views on the typology of economic structure which differ from each other:

- Reproducing; branch; hierarchical; economic (ownership structure); value;
- Material; financial and cost; the structure of demand; managment structure;
- Reproducing; territorial; branch; organizational and economic; foreign economic;
- Reproducing; branch; territorial; socio-economic; technological;

- Reproducing; branch; territorial; ownership structure; organizational and legal; the structure of investments; social; foreign economic; market infrastructure;
- Reproducing; branch; technological; regional; socio-economic; structure reflecting the process of concentration, cooperation and centralization of capital;
- Reproducing; branch; technological; spatial;
- The structure of social production; branch structure; sectoral structure (in the context of its large sectors); branch structure; the structure of production and consumption.

Each of the presented approaches is justified and certainly forms the basis for scientific discussions on the most optimal economic structuring in the country. The above views of scientists on the typology of economic structure correspond to specific conditions for the development of national and world economy and the specific period of historical development of the country and determines the differences in presented approaches.

It is also notable that most of the professional scientific literature devoted to the study of the issues of economic structure has a one-sided view of its typology, considering the economic structure only from the position of the ratio between the branches of production and in the context of reproductive processes. Also, there is still no consistency between scientists regarding a unified approach to distinguishing between different types of economic structure. The only thing that unites all scientists is the assertion that the essence of the concept 'economic structure' and its types are determined depending on the purposes of research and analysis. In our opinion, such an approach does not contribute to sustainable development of the country as a whole and the stability of the national economy in particular. This also makes it impossible to build the optimal economic structure in the country and, accordingly, reduces the effectiveness of structural shifts and their impact on the adaptation of the national economy to world trends.

The mainstream of modern development of both the world economy in general and national economies in particular is the global doctrine of sustainable development, which is closely linked with the changes (transformation) of technological economic structures and the need to ensure the global dynamic equilibrium. The term 'sustainable development' was first proposed in 1987 by the International Commission for Environmental Protection and Development. The concept was finally formed in 1992 during the UN conference in Rio de Janeiro. The essence of the concept and the term 'sustainable development' is the development that meets the needs of current generations but does not threaten the ability of future generations to meet their future needs (Report, 1987). There are two dominant approaches in the formation of this concept – biosphere-centered (the environment is not only a source of resources, but also the basis for all living things on the planet) and anthropocentric (the existence of mankind depends on the ability of future generations to meet their needs for natural resources). So, the imperative of the concept of sustainable development is the convergence of the triangle of spheres – ecology, economy, society. Considering the purely economic prerequisites for the emergence of this concept, it is worth noting the powerful influence of environmental factors on the production of the world economy. Thus, according to the International Institute for Social and Labor Studies, an increase in the concentration of greenhouse gases in the atmosphere will lead to a reduction in world production and the level of aggregate demand: if the traditional development scenario is followed, the level of production in 2030 will be 2.4% less than the current 7.2% in 2050 (as of 2012). Accordingly, destructive changes in ecological systems lead to the loss of jobs and incomes by all subjects of economic relations.

Also, one cannot ignore the influence of social factors on the development of the economy. Thus, the fundamentalists of the concept of sustainable development argue for the gradual shift of priorities in the economic and social values of man, in particular, the emphasis shifts from the purely material welfare to the non-material services (the availability of socio-humanistic services, such as quality of education, health care systems, security level, etc.) (Concerted, 2005). It provides the stimulation of higher levels of economic development. Also, a special place in the concept of sustainable development belongs to the impact of poverty on economic and environmental development. The above allows us to assert that it is impossible to form the optimal structure of the national economy without considering the concept of sustainable development and effective structural policy of the state. This makes it necessary to revise the existing approaches of scientists to the typology of economic structure. Before presenting the author's suggestions on the types of economic structures, it is advisable to consider the time factor or the life cycle of each structural type. So, the following stages of life cycle are proved in work (Krasilnykov, 2001), which pass each separate type of economic structure:

- The occurrence (a new type of structure may arise both during the formation of a new country with its economy and within the framework of already existing structure during transformation processes in the economy and, in general, by a qualitative change in the economic structure);

- The development (like the previous stage, it can occur within already existing structure);
- The maturity (the development of the structure is somewhat suspended, it acquires the features of stability, 'static' and the balance of processes occurring within the structure);
- The regressively (this period of life cycle is characterized by structural crises and imbalances that lead to a violation of the balance state of the structure and have negative consequences for its functioning);
- The decline (the structural development stops, accompanied by its replacement with a new structure).

It's worth noting that considering the stages of the life cycle presented above, which are characteristic of the economic structure, will allow timely reactions to the corresponding structural changes and deformations in the economy and processes, it is accompanied. Along with this, it is necessary to understand that within the framework of the work a generalized typology of economic structures is presented that form the basis for the study of certain aspects of the economic structure and the stages of the life cycle described above do not undergo directly suggested type of structure, but its subtypes or the state of structure that is characterized by a clearly defined period of time.

Thus, in the framework of the study of theoretical foundations of economic structure, it is proposed to improve the typology of the structure of national economy, which is shown in Figure 8.2.

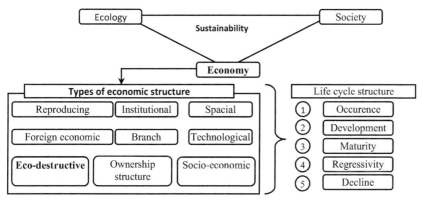

Figure 8.2 Typology of economic structure from the perspective of the concept of sustainable development of the country (author's development).

According to Figure 8.2, the author proposed a new type of structure – 'eco-destructive', which characterizes the impact of environmental losses on the activities of economic entities on the economy and sustainable development of the country as a whole. The essence of this type of structure is the distribution of the entire economy and, accordingly, production in the country to 'green' or those that prevent/do not lead to environmental pollution, provide / promote the restoration of ecological systems and rationally consume resources and those that carry out destructive influence on the environment and lead to environmental losses in the economy. Thus, ecology is a factor that determines the type of economic structure.

Along with this, it should be noted that all production in the country, regardless of industry, sector of the economy or technological equipment, should strive to reduce the volume of environmental losses. Ecological losses should be understood as "the losses in the national economy expressed in value terms (losses, additional costs, lost profits) from the eco-destructive activity of economic entities" (Melnyk, 2004). Accordingly, environmental losses lead to environmental consequences, not only in the environment, but also in the social and economic spheres, fully consistent with the fundamental principles of the concept of sustainable development.

Eco-destructive type of economic structure is characterized by deep ties between the three main components of the concept of sustainable development: the environment, social and economic spheres. This determines the need to consider the principle of synergy between these spheres for ensuring sustainable development of the country. On this basis, the traditional identification of welfare in the country and economic growth based on GDP does not meet the current requirements of the concept and does not allow for an objective assessment of these categories. That is why, within the eco-destructive type of economy, it is advisable to expand the list of indicators, except for economic processes which will consider both environmental and social. Also, within this type of structure it is important to present clearly the list of losses that can be received by the country's economy as a result of its destructive impact on the environment. Therefore, structural restructuring, as a permanent process characteristic of the Ukrainian economy, should consider the state of the environment, the need to maintain ecological balance, especially during the rapid modernization of production and innovative development and stimulate the formation of new industries. Ignoring the peculiarities of eco-destructive type of economic structure will necessarily lead to the destruction of the integrity of economic system, the loss of its quality properties and deformation of economic relations.

Summarizing this study of theoretical foundations of economic structuring, we note the need to transform the views on this concept and its main characteristics. The disclosure of substantive aspects of economic structuring should be conducted from the perspective of the concept of sustainable development, that is, considering along with social and environmental foundations. The economic structure is an indicator of its stability, sustainability, and also an indicator of ensuring the national economic security of the country as a whole. Thus, an effective economic development of the country depends not only on resource provision, the quality of ties between the subjects of economic relations, institutional environment, the level of innovations implementation, etc., but also from destructive processes in the environment. So, Ukraine changes the socio-economic paradigm for the development of the national economy. Crisis phenomena of a socio-economic nature stimulate the search for new approaches to the modernization of the national economy in different dimensions: technological, social and communication, educational, etc.

8.3 European Union Experience of Economic Restructuring

The trends of the 3rd and 4th industrial revolutions become more visible in the economic activities of the society in EU. The third industrial revolution, which is already embodied in the form of a digitalization of all spheres of life activity of society, is a prerequisite for the onset of the fourth industrial revolution and the reaction of civilization to the emergence of new opportunities for economic development within the existing socio-economic formation. Within the framework of the 4th industrial revolution, it is expected that digital technologies will have a revolutionary impact primarily on production and business. In the context of globalization transformations is important in terms of the emergence of both new advantages and risks. It should be noted that the level of risk can be significantly lower if the socio-economic systems of the country will consciously prepare and plan the necessary changes at various levels of management.

Literature analysis shows that Ukraine did not fall into the list of countries that are beneficiaries of the fourth industrial revolution. Although, according to the results of the calculations given in the analytical materials of the World Economic Forum, most countries of the world are at a low level as for the readiness of cardinal innovation changes. Only 25 developed countries that are located in Europe, North America and East Asia have high chances to implement innovative trends in economic development. In this context, it becomes clear the importance of state regulation of processes, which are designed to create the proper conditions for the timely implementation of world trends in the national economy in Ukraine con-sideling world experience.

In scientific publications on the management of socio-economic development of society, the general issues of the relevance of industrial revolutions in the context of this study are mainly considered that does not allow to form a systematic view on the issues of carrying out reforms to enhance the state economic security.

Global socio-economic transformations caused by the adoption of the UN Millennium Declaration, the implementation of the Sustainable Development Goals, the digitization of the economy in accordance with the concept of Industry 4.0, the development of alternative energy, the internationalization of economic relations and convergent processes in world markets, lead to significant changes in the structure of the national economy (national economy) of most countries. So, a striking example of the rapid restructuring of the national economy in response to global economic transformation is the experience of Sweden, where only the last decade has become dominated by high-tech industries, in particular: production volumes in the IT industry grew by 82.2%, in telecommunications – by 71%, 3%, in the automotive industry – by 15.8%, in the production of computers, electronics and optical devices – by 39.5%. In Switzerland, the development of green and energy-saving technologies has led to the predominance of the services sector (in 2016, 75.3% of the economically active population is employed in the field) over the production sphere.

8.4 Methodological Approach to the Evaluation of the Potential of Types of Economic Activities

The formation of scientific and methodological support for assessing the potential of types of economic activities considering the environmental component requires the study of already existing mathematical modeling tools of this category. Thus, while carrying out a critical analysis of proposed approaches to assessing the potential [10–12], it should be noted that, in parallel with the use of sufficiently powerful mathematical tools, in view of the priority of indicators of the attribute space, the use of a convolution of comprehensive directions of the characteristics of the research object in a generalizing, the failure to take account of economic nature of the potential, that is, a certain comparison of actual achieved level and the optimal reference level remains one of the problematic aspects.

The nature of the proposed method is to assess the potential of types of economic activities given the environmental component based on the deviation of normalized indicators of attribute space from the values of the "reference" level. So, following quasi-distances, a quantitative assessment of the potential of each type of economic activity is made, considering the environmental component, which makes it possible to establish an unachieved level in comparison with possible 100%. Turning to the adaptation of taxonometric method that satisfies all the above conditions, to assess the potential of economic activities subject to the environmental component, let's consider the following sequence of its formalization:

1. Formation of a set of indicators of attribute space, reflecting the potential of types of economic activities. To visualize this process, it is proposed to build a table layout that characterizes the values of indicators of attribute space, constructed on the basis of statistical reporting in the context of each type of economic activity.
2. Calculation of matrix values of indicators of the attribute space, reflecting the potential of types of economic activities considering the environmental component. This matrix acts as a concentrated expression of information describing different types of economic activity in the context of economic, social and environmental components.
3. Carrying out the normalization (bringing into comparable form) defined in points 1 and 2 of the indicator system of attribute space, which implies their transformation into a comparable form.
4. Formation of "reference" values characterizing economic, social and environmental components of types of economic activities.
5. Calculation of quasi-distances based on comparison of normalized indices of attribute space, characterizing the potential of types of economic activities, considering the environmental component with similar values of "reference" level.

To visualize the calculation procedure of taxonomic methods for assessing the potential of types of economic activities considering the environmental component, authors will perform the mathematical formalization of the calculation stages for solving the assigned task of economic and mathematical modeling.

At the first stage the identification of indicators of the attribute space characterizing the potential of types of economic activities considering the environmental component takes place. The indicators represented in Table 8.1 become the input informative base for constructing a given set of coefficients.

Continuing the formalization of a given taxonometric approach to assessing the potential of types of economic activities and considering the environmental component (*the second stage*), it is proposed to present the totality of considered indicators of the attribute space in a matrix form (Equation 8.1).

$$K = \begin{pmatrix} k_{11} & \cdots & k_{1j} & \cdots & k_{1n} \\ \cdots & \cdots & \cdots & \cdots & \cdots \\ k_{i1} & \cdots & k_{ij} & \cdots & k_{in} \\ \cdots & \cdots & \cdots & \cdots & \cdots \\ k_{m1} & \cdots & k_{mj} & \cdots & k_{mn} \end{pmatrix}$$

$$= \begin{pmatrix} k_{11} & k_{12} & k_{13} & k_{14} & k_{15} & k_{16} \\ k_{21} & k_{22} & k_{23} & k_{24} & k_{25} & k_{26} \\ k_{31} & k_{32} & k_{33} & k_{34} & k_{35} & k_{36} \\ k_{41} & k_{42} & k_{43} & k_{44} & k_{45} & k_{46} \\ k_{51} & k_{52} & k_{53} & k_{54} & k_{55} & k_{56} \end{pmatrix} \quad (8.1)$$

where K – the matrix of indices of attribute space characterizing the potential of types of economic activities considering the environmental component; $i = 1 \div m$ – number of corresponding type of economic activity; $j = 1 \div n$ – number of corresponding indicator of attribute space; k_{ij} – the j-th indicator evaluating the potential of the i-th type of economic activity.

This approach to the presentation of incoming statistical data on the assessment of the potential of types of economic activities taking into account the environmental component allows both to concentrate and visualize the informative base of the model and simplify the calculation procedure based on matrix operations.

At the *third* stage of implementation of taxonomic approach to definition the potential of types of economic activities considering the environmental component. It is advisable to bring the indicators of attributive space into a comparable form (Equation 8.2):

$$L = \begin{pmatrix} l_{11} & \cdots & l_{1j} & \cdots & l_{1n} \\ \cdots & \cdots & \cdots & \cdots & \cdots \\ l_{i1} & \cdots & l_{ij} & \cdots & l_{in} \\ \cdots & \cdots & \cdots & \cdots & \cdots \\ l_{m1} & \cdots & l_{mj} & \cdots & l_{mn} \end{pmatrix} = \begin{pmatrix} l_{11} & l_{12} & l_{13} & l_{14} & l_{15} & l_{16} \\ l_{21} & l_{22} & l_{23} & l_{24} & l_{25} & l_{26} \\ l_{31} & l_{32} & l_{33} & l_{34} & l_{35} & l_{36} \\ l_{41} & l_{42} & l_{43} & l_{44} & l_{45} & l_{46} \\ l_{51} & l_{52} & l_{53} & l_{54} & l_{55} & l_{56} \end{pmatrix}$$

$$(8.2)$$

Table 8.2 Indicators characterizing the potential of types of economic activities, considering the environmental component in 2015

Types of Economic Activity	Economic Component		Social Component		Environmental Component	
	EC1	EC2	SC1	SC2	EnC1	EnC2
Agriculture, forestry and fishery	558788	27900	1.2	119.7	77.7	8736.8
Mineral industry and quarry development	186194	17246.3	0.2	6.3	460.9	232642.4
Process industry	1206047	44563.1	5.1	73.7	941.4	56506.3
Supply of electricity, gas, steam and conditioned air	176768	21039.9	22.3	385.1	1174.3	6597.5
Water supply; drain system, waste management	26982	1318.7	2.4	40.8	9	594.2
Building industry	188595	40931.5	1	12.2	3.4	89.9
Wholesale and retail trade; repair of vehicles and motorcycles	549163	18152.4	4.1	71.4	4.82	31.42
Transport, warehousing, postal and courier activities	295634	16278	2.8	19.9	72.32	471.24
Temporary accommodation and arrangements for feeding	25458	970	0.8	13.3	24.11	157.08
Information and telecommunications	142223	21848.4	0.3	3.6	4.02	26.18
Financial and insurance activities	107764	6223.7	0.2	4.2	2.41	15.71
Real estate	176078	8797.6	0.4	4.2	1.61	10.47

(*Continued*)

Table 8.2   Continued

Types of Economic Activity	Economic Component		Social Component		Environmental Component	
	EC1	EC2	SC1	SC2	EnC1	EnC2
Professional, scientific and technical activities	107124	3805.3	0.6	7.8	0.80	5.24
Activities in administrative and support services	43370	5677.4	1.1	8.1	6.43	41.89
Public administration and defense; compulsory social insurance	147578	12547.5	2.7	26.9	33.75	219.91
Education	119928	1176.4	1.4	27.2	0.96	6.28
Health care and social assistance	88636	1550.2	2.1	19.3	2.25	14.66
Arts, sports, entertainment and recreation	20436	921.4	0.3	2.8	3.21	20.94
Providing other services	22475	206.5	0.4	5.3	4.02	26.18

Notes: EC1 – The output of products and services in actual prices; mln. UAH; EC2 – Capital investments (used) in actual prices, mln. UAH; SC1 – Employers' need for workers, ths. people; SC2 – Work of registered unemployed, ths. people; EnC1 – Air pollutant emissions by stationary sources of pollution, ths. tons; EnC2 – Formation of waste of I–IV hazard classes, ths. tons.

that is, normalizing indicators to a comparable form, which is proposed to be carried out with the help of equations:

$$l_{ij} = \left| \frac{k_{ij} - \overline{k_i}}{\sigma_i} \right|$$

$$\overline{k_i} = \frac{1}{m} \sum_{j=1}^{n} k_{ij}, \sigma_i = \frac{1}{m} \sum_{j=1}^{n} (k_{ij} - \overline{k_i})^2, \tag{8.3}$$

where L – the matrix of normalized indicators of attribute space characterizing the potential of types of economic activities considering the environmental component;

l_{ij} – normalized j-th indicator of evaluation of potential of the i-th type of economic activity.

$\overline{k_i}$ – the average value of the j-th indicator for the totality of considered types of economic activities;

σ_i – standard deviation of the j-th index.

The need for the normalization of indicators of attribute space is due to the heterogeneity of forms of expression and the units of measurement in which they are represented.

At the *fourth stage* the calculation of "reference" values of indicators of attribute space, from the point of view of assessment of potential of types of economic activity considering the environmental component which is proposed to present as the matrix

$$
\begin{pmatrix}
l_{11et} & \cdots & l_{1jet} & \cdots & l_{1net} \\
\cdots & \cdots & \cdots & \cdots & \cdots \\
l_{i1et} & \cdots & l_{ijet} & \cdots & l_{inet} \\
\cdots & \cdots & \cdots & \cdots & \cdots \\
l_{m1et} & \cdots & l_{mjet} & \cdots & l_{mnet}
\end{pmatrix},
$$

where l_{ijet} – normalized j-th reference indicator is performed.

The determination of the reference normalized value of the j-th indicator of attribute space for the whole set of considered types of economic activity (maximum or minimum value depending on the direction of the impact on the outcome). Thus, the "reference" value of a corresponding indicator of attribute space is calculated as a minimum value, if an increase in this indicator leads to a loss of a certain level of potential, and the maximum value – otherwise. The maximum value for the model under consideration is selected for indicators characterizing the economic and social components of the potential.

Calculations of normalized "reference" values of indicators of attribute space are made by adjusting the optimal normalized value of the j-th indicator of potential of types of economic activities by the magnitude of standard deviation (Equation 8.4), defined in clause 4.1:

$$
L_{et} = \begin{pmatrix}
\max_j\{k_{1j}\} + \sigma_j & \cdots & \min_j\{k_{1j}\} + \sigma_j & \cdots \\
\max_j\{k_{1j}\} + \sigma_j & \cdots & \min_j\{k_{1j}\} + \sigma_j & \cdots \\
\max_j\{k_{1j}\} + \sigma_j & \cdots & \min_j\{k_{1j}\} + \sigma_j & \cdots \\
\max_j\{k_{1j}\} + \sigma_j & \cdots & \min_j\{k_{1j}\} + \sigma_j & \cdots \\
\max_j\{k_{1j}\} + \sigma_j & \cdots & \min_j\{k_{1j}\} + \sigma_j & \cdots
\end{pmatrix}, \qquad (8.4)
$$

where L_{et} – the matrix of normalized "reference" values of indicators of the attribute space; $\max_j\{k_{1j}\}$ (accordingly $\min_j\{k_{1j}\}$) – the determination of maximum (respectively, minimum) normalized value of the j-th indicator, depending on the direction of their influence on the outcome.

The basis for adjusting the reference values of indicators of attribute space by the average deviation is the assumption of the impossibility to achieve the "reference". Considering this fact, it can be noted that the "reference" is the value larger than the optimal normalized value of the j-th indicator of attribute space. That is why, in our case, it is advisable to use the standard deviation as the correction coefficient, which characterizes the remoteness of the average value of the indicator from individual levels.

This stage is the basis of taxonomic approach to assessing the potential of types of economic activities taking into account the environmental component, since the "reference" level from the point of view of the potential provides an opportunity to form a tactics and development strategy, bringing the attribute space values to the reference value.

The *fifth stage* of scientific and methodological approach to the assessment of potential of types of economic activities considering the environmental component provides the calculation of quasi-distances between the i-th and "reference" levels based on the following equation:

$$KV_i = \sum_{j=1}^{n-2} (l_{ij} - l_{ijet})^2 - \sum_{j=n-1}^{n} (l_{ij} - l_{ijet})^2, \tag{8.5}$$

where KV_i – quasi-distance between the i-th and "reference" levels.

The quasi-distances are calculated and presented in Table 8.3, and act as a quantitative characteristic, which shows the correspondence of the level of potential of economic activities considering the environmental component of the reference value.

Based on the data given in Table 8.4, it is fair to note that in Ukraine most of the economic activities are not realized even by half. Such types of economic activities as transport, warehousing, postal and courier activities, public administration and defense; compulsory social insurance have the potential opportunities for growth at a level of more than 63%. If it can be argued on transport, that a significant part of the realization of its potential lies within the environmental component, namely the minimization of emissions into the atmosphere from motor vehicles, then the state administration and defense, its potential capacities accumulate within the economic and social component.

The lowest value of the potential, considering the environmental component, has a supply of electricity, gas, steam and conditioned air and makes

Table 8.3 Normalized indicators (the first line for each type of economic activity) and quasi-distances (the second line for each type of economic activity), which characterize the potential of types of economic activities taking into account the environment all component

Types of Economic Activity	Economic Component		Social Component		Environmental Component	
	EC1	EC2	SC1	SC2	EnC1	EnC2
Agriculture, forestry and fishery	1.189	1.091	0.281	0.851	0.210	0.137
Mineral industry and quarry development	9.179	3.271	20.593	14.680	0.539	0.735
Process industry	0.121	0.299	0.482	0.438	0.920	4.009
Supply of electricity, gas, steam and conditioned air	16.798	6.762	18.809	18.013	0.001	9.091
Water supply; drain system, waste management	3.465	2.329	0.502	0.328	2.337	0.748
Building industry	0.568	0.325	18.635	18.957	1.942	0.061
Wholesale and retail trade; repair of vehicles and motorcycles	0.154	0.581	3.967	3.866	3.024	0.176
Transport, warehousing, postal and courier activities	16.527	5.375	0.727	0.666	4.328	0.669
Temporary accommodation and arrangements for feeding	0.680	0.884	0.047	0.046	0.412	0.287
Information and telecommunications	12.523	4.061	22.771	21.489	0.282	0.499
Financial and insurance activities	0.112	2.059	0.322	0.371	0.429	0.297
Real estate	16.867	0.706	20.230	18.587	0.265	0.486

(*Continued*)

Table 8.3 Continued

Types of Economic Activity	Economic Component		Social Component		Environmental Component	
	EC1	EC2	SC1	SC2	EnC1	EnC2
Professional, scientific and technical activities	1.156	0.367	0.301	0.302	0.425	0.298
Activities in administrative and support services	9.385	6.416	20.411	19.185	0.269	0.485
Public administration and defense; compulsory social insurance	0.264	0.227	0.040	0.283	0.226	0.290
Education	15.641	7.141	22.840	19.349	0.515	0.496
Health care and social assistance	0.686	0.910	0.362	0.358	0.368	0.296
Arts, sports, entertainment and recreation	12.485	3.958	19.870	18.695	0.332	0.488
Providing other services	0.275	0.641	0.462	0.468	0.427	0.298

Notes: EC1 – The output of products and services in actual prices; mln. UAH; EC2 – Capital investments (used) in actual prices, mln. UAH; SC1 – Employers' need for workers, ths. people; SC2 – Work of registered unemployed, ths. people; EnC1 – Air pollutant emissions by stationary sources of pollution, ths. tons; EnC2 – Formation of waste of I–IV hazard classes, ths. tons.

18.3%. This is quite logical, since the supply of electricity and gas is relatively environmentally friendly but to it, even in Ukraine, conditions are set in terms of minimizing environmental risks. Assessing their opportunities within the economic and social components, it is fair to note that currently there are no vectors for the development of these types of economic activities. Power lines and gas pipelines cover almost all the necessary territory of Ukraine, and there are no investment resources for improvement in sufficient volume. The potential for all other types of economic activity in Ukraine are in the range of values from 47% to 60% and is formed from all three of its components. So, there are more than enough opportunities for restructuring Ukraine's

Table 8.4 The potential of types of economic activities, taking into account the environmental component

Types of Economic Activities	Potential
Agriculture, forestry and fishery	46.45
Mineral industry and quarry development	51.29
Process industry	36.48
Supply of electricity, gas, steam and conditioned air	18.30
Water supply; drain system, waste management	60.06
Building industry	55.64
Wholesale and retail trade; repair of vehicles and motorcycles	54.64
Transport, warehousing, postal and courier activities	63.96
Temporary accommodation and arrangements for feeding	54.19
Information and telecommunications	56.64
Financial and insurance activities	56.15
Real estate	59.35
Professional, scientific and technical activities	56.37
Activities in administrative and support services	56.24
Public administration and defense; compulsory social insurance	66.06
Education	59.26
Health care and social assistance	59.05
Arts, sports, entertainment and recreation	52.22
Providing other services	52.47

economy, each type of economic activity can act as an impulse and a leader in structural changes.

8.5 Conclusions

Over the past decades the Ukrainian economy seeks to transform in reply to new challenges in the home country and globalization processes abroad. Many development programs and strategic plans for the restructuring of political, social, environmental and economic areas of the country are established almost every year and at different times, including a significant list of goals and objectives, which should help Ukraine to achieve sustainable development. However, economy of our country is still raw materials oriented, resource-intensive, energy-inefficient, and eighty percent of wear and tear and extremely unclean production. A dynamic study of technogenic pollution of environment and volume of industrial wastes in Ukraine provide evidences of catastrophic ecological effects on the commonwealth.

The above actualizes the formation of public policy in greening of the Ukrainian economy not only with a clear theoretical formalization of its implementation stages, but also an accurate understanding of quantitative indexes which should be achieved at each stage of economic restructuring. Thus, the development of the assessment of the level of potential of types of economic activities taking into account the environmental component, which, unlike the existing ones, is proposed to be viewed as a combination of three components (economic, social and environmental) through a transition to normalized indicators (weighing the deviation of a current level by average on the standard deviation) and their subsequent convolution to a single integral criterion – quasi-distances based on the deviation from the reference level, calculated by adjusting the maximum (for stimulants) and the minimum (for dissimulators) values by the value of standard deviation, which allows to establish unachieved level as compared with possible 100%.

Acknowledgements

This chapter was prepared as a part of the Scientific Project "Modeling the Transfer of Eco-Innovations in the Enterprise-Region-State System: Impact on Ukraine's Economic Growth and Security" (No. 0119U100364), that was financed by the state budget of Ukraine.

References

Alekseiev, I. V., Kolisnyk, M. K. and Moroz, A. S. (2007). Upravlinnya resursnym zabezpechennyam promyslovo-finansovykh hrup: [monohrafiya] [Management of resource support of industrial and financial groups: [monograph]. Lviv: Publishing House of Lviv Polytechnic National University, 132 p.

Ananidze, V. Ya (1993). Sushchnost' strukturnoy perestroyki, formy i metody yeye osushchestvleniya [The essence of structural adjustment, forms and methods of its implementation]. In Innovatsionnaya i investitsionnaya politika strukturnoy perestroyki narodnogo khozyaystva [Innovative and investment policy of structural reorganization of the national economy]. Moscow, pp. 143–156.

Bell, D. (1999). Gryadushcheye postindustrial'noye obshchestvo [The Future of Post-Industrial Society]. In Opyt sotsial'nogo prognozirovaniya. [The Experience of Social Prediction]. M.: Academia, 956 p.

Benson, Emily and Oliver Greenfield (2016). Inclusive Green Economy. Green and Fair. UNON, Publishing Services Section.

Bublyk, Maryna B. and Maria R. Bei (2016). Features of the "Green" economy and key tools for its transformation in a social-oriented system. Bulletin of the Lviv Polytechnic National University. Problems of Economics and Management, 847: 29–34.

Burkinsky, B. V., Halushkina, T. P. and Ye V. Reutov (2011). "Green" economy in light of the transformational changes in Ukraine. Phenix Enterprise.

Burych, I. V. (2016). Forming a portfolio of innovative projects to provide "green" economic growth in the region. PhD diss., Sumy State University.

Bystryakov, Igor K. (2011). Establishment the "green" economy in Ukraine: methodological aspects. Mechanism of economic regulation, 4: 50–57.

Cato, M. S. (2009). Green Economics: An Introduction to Theory, Policy and Practice. Earthscan.

Clark, C. The Conditions of Economic Progress / Clark Colin. – London: Macmillan, 1940. 504 p.

Concerted Development of Social Cohesion Indicators. – Strasbourg: Council of Europe Publishing, 2005. 235 p.

Dzagoieva, I. T. and Tskhurbaeva, F. H. (2008). Gosudarstvennaya politika formirovaniya perspektivnoy struktury ekonomiki regional'nogo APK [State policy of formation of perspective structure of economy of regional agroindustrial complex]. Vestnik of Rostov State University of Economics, 4/2(6), 365–368.

Fourastié, J. (1951). Le progrès technique et l'évolution économique/J. Fourastié// Institut d'Études Politiques de Paris. – Paris, les cours de Droit (deux fascicules), 249 p.

Klimova, O. I. (2015). Struktura ekonomichnoii systemy: terminolohichnyi analiz [Structure of the economic system: terminological analysis]. Molodyi vchenyi, 2(17), 1112–1115.

Krasilnykov, O. Yu (2001). Strukturnyye sdvigi v ekonomike [Structural shifts in the economy]. Saratov, SSU, 160 p.

Maliuk, V. I. and Nemchyn, A. M. (2008). Proizvodstvennyy menedzhment: Uchebnoiye posobiye [Production Management: Textbook]. SPb: Peter, 288 p.

Melnyk, Leonid G. (2016). Means and key factors of sustainable ("green") economy formation. Actual problems of economics, 178: 30–36.

Melnyk, L. H. and Karintseva, O. I. (eds.) (2004). Metody otsinky ekolohich-nykh vtrat: monohrafiya [Methods of estimation of ecological losses: monograph]. Sumy, 288 p.

Potapenko and Vyacheslav H. (2012). Strategic priorities of safety development of Ukraine on the principles of "green economy". NISS

Prushkivs'ka, E. V. (2013). Evolyutsiia kontseptsii strukturuvannia natsional'noii ekonomiky [Evolution of the concepts of structuring the national economy]. Problems of the Economy, 2: 87–94.

Prushkivska, Emiliya V. and Yulia O. Shevchenko (2013). "Green economy" development: national aspect. Business-inform, 3: 186–191.

Report of the World Commission on Environment and Development: Our Common Future/Transmitted to the General Assembly as an Annex to document A/42/427 – Development and International Co-operation: Environment. – Oxford: Oxford University Press, 1987. 300 p.

Reshetylo, V. P. (ed.) (2009). Natsional'na ekonomika: navch. posibnyk [National economy: textbook]. Kharkiv: O. M. Beketov NUUE, 386 p.

Romanenko, M. A., Yemets, M. A. Romanenko, I. I. and Petrovtsij, O. V. (2016). European integration as a factor in accelerating the implementation of the foundations of the "green" economy in Ukraine. Construction. Material science. Mechanical engineering. Series: Power engineering, ecology, computer technologies in building, 92: 114–119.

Sotnyk, Iryna M. (2016). Energy efficiency of Ukrainian economy: problems and prospects of achievement with the help of ESCOs. Actual problems of economics, 1(175): 192–199.

The World Bank [Electronic resource]. Access mode: http://data.worldbank.org.

9

Renewable Energy to Overcome the Disparities in Energy Development in Ukraine and Worldwide

Iryna Sotnyk[1,*], Mykola Sotnyk[2] and Iryna Dehtyarova[1]

[1]Department of Economics, Entrepreneurship and Business Administration,
Sumy State University, Ukraine
[2]Research Institute of Energy Efficient Technologies,
Sumy State University, Ukraine
E-mail: insotnik@gmail.com
*Corresponding Author

9.1 Introduction

The need for energy supply for all nations is basic and constantly growing in the modern world. This is due to the development of scientific and techno-logical progress and the implementation of its achievements in management practices accompanied by the increase in the amount of energy needed for people's everyday life to meet the ever-increasing living standards. The unequal territorial distribution of non-renewable fossil fuels, which are the basis of energy complexes in most countries, highlighted the search for traditional energy resources that would be sufficient to meet the needs of national economies. The resource scarcity in many countries leads to political and military conflicts in the struggle to get the access to energy sources, slows down economic growth of nations, highlights social inequality and causes the emergence of "energy poverty" phenomenon (EU Energy, 2018; Zavgorodnya, 2017). In addition to social, economic and political negative impacts increased consumption of fossil fuels results in environmental pol-lution increase, breaking the balance of ecosystems and causing the threat of global warming.

The 70s and 80s of the 20th century energy crises, caused by the lack of available fossil fuels, clearly demonstrated the need of the world community to move towards renewable energy sources (RES). Its potential is practically unlimited at this stage in the development of productive forces and can be exploited by each country. In order to implement national sustainable development strategies and ensure their own energy security, the world's leading developed countries as well as developing countries have been actively developing the renewable energy (RE) sector for the last 40 years.

The economic, social, political, and environmental benefits of RE draw the government's attention to the issue of deploying green power capacities. Mechanisms for RE management and technology development are the research subject for many scholars. However, the practical implementation of alternative energy projects often faces organizational, economic, social and other difficulties and requires strong state support for their solution. In this regard, this study examines the preconditions, trends and mechanisms of RE development worldwide and in Ukraine, analyzes the issues of alternative energy sector development for domestic households as the youngest RE market participant. Following the SWOT analysis some key measures and considerations to overcome the disparities in energy development in Ukraine due to RE progress to achieve sustainable development goals may be considered.

9.2 Preconditions, Trends and Mechanisms for Renewable Energy Development in the World

Today, governments consider RE as a real opportunity to reduce the scarcity and intermittence of fossil fuels worldwide, stop energy resource wars, increase employment and life quality, and decarbonize national economic systems. The social aspect of RE development for achieving sustainable development goals is constantly tackled in the international and national documents related to the development of this sector. The advantages of the implementation of modern RE technologies include energy supply for the population living in remote and inaccessible areas, as well as access to electricity for the poorest people: for example, the Solar Home System program, which helps the rural population of one of the poorest countries in the world Bangladesh to install solar systems and receive electricity (Dhaka Tribune, 2017). In addition, the alternative energy sector is gradually becoming a powerful employer, creating new jobs; providing health safety; not causing problems related to waste utilization and emission of harmful substances and it is not a terror attack target. Its development allows investing in the local economy and increasing the living standards of the population (Chernyak and Fareniuk, 2015).

Wind energy, solar, geothermal, bio-hydropower and environmental energy (heat pumps) are among the most demanded technologies in the world today (U.S. EIA, 2013). Their use depends on the climatic conditions (solar, wind energy), the availability of sufficient raw material (biomass energy), specific natural conditions (geothermal energy), water resources (hydropower, environmental energy), etc. At the same time, even under favorable natural conditions for RE development, such technologies are currently subsidized; they require governmental economic support. The reason is poor technology for renewable energy generation, that determines its high cost, and, consequently, long payback periods for projects, that deter wide and fast distribution of green power capacities. Kurbatova (2016) indicates low energy density, its intermittence on the Earth's surface (in terms of hours, days of the year, geographic zones) as the most significant disadvantage of RE as well as high initial capital expenditures, though usually compensated by low operating costs, however, significantly influence alternative energy generation costs.

Consequently, most countries that develop RE use economic incentives for its development. The most highly used incentives for RE deployment in private and business sectors are (Klopov, 2016; Kurbatova, 2016; Kurbatova and Khlyap, 2015; Kurbatova et al., 2014; Riazanova, 2017):

- Direct incentives – financial incentives for RE producers, implemented through the use of certain economic mechanisms (preferential tariffs, preferential premiums, green certificates, tender schemes, investment grants, tax and customs privileges, subsidies, bonuses, etc.);
- Indirect incentives – encourage the use of RES directly by reducing the attractiveness of fossil energy resources through the introduction of environmental taxes, CO_2 tax, etc.;
- Voluntary programs that foresee consumers' will of to pay high energy prices for renewable energy sources through environmental care in order to maintain long lasting stability. Such programs include donation programs and charity projects.

Economic incentives in RE sector showed almost 6 times growth of world investments in alternative energy in the last 13 years: from \$47 (2004) to \$279.8 (2017) billion (Figure 9.1). At the same time, the highest investment period was in 2015, with its maximum of \$323.4 billion. It is remarkable that in 2015, for the first time in history, developing countries spent more

Growth:

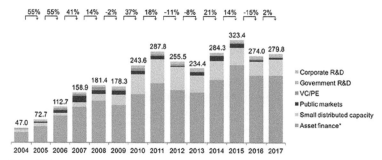

Figure 9.1   Global new investment in renewable energy by asset class, 2004–2017, $ billion (UN Environment, 2018).

*Asset finance volume adjusts for re-invested equity. Total values include estimates for undisclosed deals.
Source: UN Environment, Bloomberg New Energy Finance.

on renewable energy than developed countries. In 2016–2017 this trend was preserved. In 2017, the gap grew sharply, so that the developing world accounted for 63% of the global total and developed countries just 37% (Figure 9.2). In absolute numbers developing economies (including China, Brazil and India) committed $177 billion to renewables, up 20%, compared to $103 billion for developed countries, down 19%. This was the largest tilt in favor of developing countries yet seen (UN Environment, 2018). Thus, the myth that alternative energy is the prerogative of wealthy countries has been refuted.

In general, net renewable capacity change as a percentage of global capacity change increased from just under 20% in 2007, to 39% in 2013, to 57% in 2016 and 61% in 2017. Currently the world increases renewable capacity faster than traditional one, and if such trend continues, in the long run, we should expect a global transition to energy production mostly from RES. RE progresses both in terms of installed capacity and power generation. So, over the past decade, the share of renewable capacity in global power capacity increased from 7.5% in 2007 to 19% in 2017, more than 2.5 times. Instead, in 2017, RE share of total electricity produced grew to 12.1%, up from 5.2% in 2007, i.e. 2.33 times (UN Environment, 2018).

The additional argument for the deployment of RES capacities is RE technology markets, which demonstrate the tendency to reduce alternative energy generation cost. The largest price breakthrough is typical for solar power. So, between 2009 and 2017, the benchmark levelized cost of electricity, or LCOE, for photovoltaics without tracking systems fell from $304 per megawatt-hour

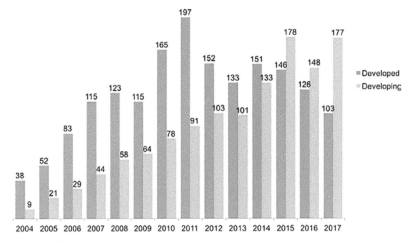

Figure 9.2 Global new investment in renewable energy: developed v. developing countries, 2004-2017, $ billion (UN Environment, 2018).

New investment volume adjusts for re-invested equity. Total values include estimates for undisclosed deals. Developed volumes are based on OECD countries excluding Mexico, Chile, and Turkey.
Source: UN Environment, Bloomberg New Energy Finance.

to just $86, a reduction of 72%. Some of this was due to a fall in capital costs, some to improvements in efficiency. Exept solar energy cost, onshore wind's LCOE dropped from $93 to $67 per MWh, a reduction of 27%. For offshore wind, there was an increasing cost trend for some years as project developers moved into deeper waters, further from shore, but since the peak in 2012, there has been an LCOE decline of 44% to $124 per MWh (UN Environment, 2018). Bloomberg New Energy Finance forecasts cost of electricity generated by wind and solar power plants will fall by 41% and 60% accordingly by 2040 and in 25 years they will become the cheapest means of energy production. The capacity of the world's marine wind power will increase 6.5 times to 114.9 GW by 2030 that means the sector will grow 16% per year on average (Association, 2018).

One of the ways to overcome RE disadvantages, intermittent character of alternative energy generation due to the high dependence on changes in natural conditions (day hours, weather, etc.), is to create the systems for the accumulation and storage of electricity from RES generators. Until recently, such systems were quite costly, but in recent years the situation has changed for the better: in the years 2010–2017, the average cost of energy storage systems (ESS) with lithium-ion batteries decreased 4.78 times – from $1,000 (2010) to $209 (2017) per kW (Figure 9.3), allowing potential consumers

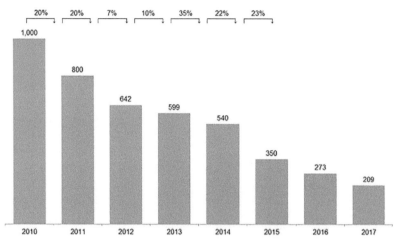

Figure 9.3   Lithium-ion battery pack price, global average, $/kWh (UN Environment, 2018).
Source: UN Environment, Bloomberg New Energy Finance.

more access to alternative energy. In addition, RES contributed to increased energy supply reliability.

An important social aspect for RE development is its gradual transformation into a considerable employer. Thus, in 2017 alternative energy sector including large hydropower employed 10.3 million people, directly and indirectly. This represents an increase of 5.3% over the number reported the previous year. Employment trends in RE sector are shaped by its distribution around the world: despite the fact that alternative energy is developing in many countries, the main employers in the sector are a handful of countries, with China, Brazil, the United States, India, Germany and Japan in the lead. China alone accounts for 43% of all renewable energy jobs. Its share is particularly high in solar heating and cooling (83%) and in the solar photovoltaic (PV) sector (66%), and less so in wind power (44%) (IRENA, 2018).

In 2017, the PV industry was the largest employer (almost 3.4 million jobs, up 9% from 2016). Biofuels employment (at close to 2 million jobs) expanded by 12%, as production of ethanol and biodiesel expanded in most of the major producers. Brazil, the United States, the European Union and Southeast Asian countries were among the largest employers. Employment in wind power (1.1 million jobs) and in solar heating and cooling (807 000 jobs) declined as the pace of new capacity additions slowed. Large hydropower employed 1.5 million people directly, of whom 63% worked in operation and

Growth:

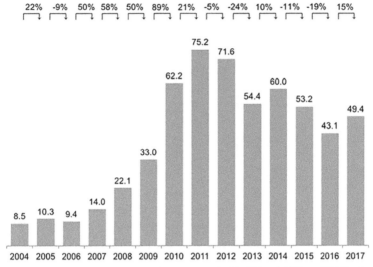

Figure 9.4   Small distributed capacity investment, 2004–2017, $ billion (UN Environment, 2018).

Represents investments in solar PV projects with capacities below 1 MW.
Source: Bloomberg New Energy Finance.

maintenance. Key job markets were China, India and Brazil, followed by the Russian Federation, Pakistan, Indonesia, Iran and Viet Nam (IRENA, 2018).

A promising trend in terms of improving access of the population to high-quality energy supply is the development of the small distributed capacity, which mostly refers to solar energy capacities. It includes a mix of residential and commercial rooftop systems, off-grid units in remote areas, and ground-mounted projects with tens, hundreds or a small number of thousands of panels. In the last decade, small distributed capacity is experiencing an investment boom: during 2010–2017, annual investments in this sector steadily exceeded the annual amounts invested before 2010 (Figure 9.4). Consequently, between 2010 and 2015, the average invested in small-scale renewables was $63 billion a year, with a peak of $75.2 billion in 2011 (UN Environment, 2018). During 2016–2017, investment declined somewhat, but still remains at a higher level than in the previous decade.

The attractiveness of investing in small distributed capacity is growing due to falling prices for solar modules used in residential sector. Thus, in 2010, the average cost per Watt of a residential PV system in Germany was

$3.90, but by the end of 2017 it had fallen 57% to $1.68. In Australia, a 4 kW PV system cost an average of $6.40 per Watt in 2010, yet by the end of 2017 it had plummeted 78% to just $1.40 per Watt (UN Environment, 2018). Recent research of the global RE technology market shows that in 2018 and beyond the prices will continue to fall, encouraging households around the world to build private low power RE plants to generate energy at an affordable price.

Consequently, it is worth forecasting further deployment of RE capacities both in business and residential sectors in different countries of the world considering the declining price trends in RE technology markets, the improvement of alternative energy technology, and creating new jobs. At the same time, we consider that households need more active state support, because low paying capacity prevents from investing sufficient financial resources in RE technology, and loans are usually expensive and payback time for such projects is too long. Nevertheless, it is advisable to develop such initiative at the state level in order to create competition in the energy market between large and small producers of "green" and "brown" electricity. It will ensure prerequisites for lowering electricity prices, raising the levels of self-sufficiency of countries on the basis of alternative energy development, and wider public access to electricity. All these measures will contribute to more sustainable territorial development, reduction of energy supply disparities, social, economic and environmental improvements.

9.3 Alternative Energy Development in Ukraine

The issues of reliable energy supply are extremely relevant for Ukraine, which has a shortage of its own energy resources and is forced to import more than one third of the necessary energy resources. The country has a high energy intensity of the gross domestic product, which is 2 times higher than the average world index and 3 times higher than the average European index (World Energy Council, 2018). For several decades Ukraine cannot get out of the "energy hole", spending billions of hryvnia to pay for expensive and scarce energy resources consumed extremely inefficiently by domestic businesses and households. The domestic energy complex is based on the outdated technology of using non-renewable energy sources, first of all, coal and natural gas, as well as atomic energy, while the share of RES in the country's electric balance does not exceed 1.47% (at the end of 2017) (NCSREPU, 2018).

At the same time, the technically feasible potential of RE in Ukraine is over 98.0 million tons of conventional fuel per year. The experts prove to refuse energy resources import if RE potential is realized alongside with energy efficiency improvements of the national economy. RE potential structure is the following: 28.5% falls on wind, 6% for solar power, 31.6% for bioenergy, 11% for geothermal heat energy and 18.4% for environment energy (heat pumps) (SAEEESU, 2019). In addition, despite the legally recognized RES development priority and the international commitment made by Ukraine to achieve 11% of renewable energy share by 2020 (CMU, 2014) in the overall energy consumption of the country, alternative energy potential is realized much more slowly than governmental programs and plans declare.

In an effort to achieve high rates of RE development, the government of the country has created almost greenhouse conditions for the development of alternative energy generation. In the last decade, many incentives were introduced to support the acceleration of renewable energy capacities in private and business sectors (CMU, 2014):

– Reduction of land tax for RE enterprises;
– Exemption from taxation: (1) profit from the main activities of companies in energy sector, which produce electricity from renewable energy sources; (2) profit of biofuel producers from biofuel sales; (3) profit from the simultaneous production of electric and thermal energy and/or the production of thermal energy with the use of biological fuels; (4) producers' profits from machinery, equipment and facilities for the manufacture and reconstruction of technical and transport vehicles consuming biological fuels;
– Exemption from the value added tax for import operations into the customs territory of Ukraine for RES equipment, equipment and materials for the production of alternative fuels or energy from RES, exemption from import duty on such equipment, facilities and materials;
– Feed-in tariffs for electric energy generated by power stations from alternative energy sources (except blast furnace and coke gases, but with the use of hydropower – produced only by micro, mini and small hydroelectric power stations) (The Verkhovna, 2017);
– Feed-in tariff premium for the use of equipment of Ukrainian origin (The Verkhovna, 2018).

Practically, feed-in tariffs, introduced by the government in 2009, turned out to be the most effective economic instrument. Today they are the highest in Europe and allow Ukrainian RES owners benefiting from alternative energy (NICU, 2018). According to the current Ukrainian legislation, feed-in tariff is valid until December 31, 2029, and its rates are differentiated, depending on commissioning and types of RES plants. They include solar, wind, small (including micro and mini) hydroelectric power plants, biomass and biogas power plants, and power plants that use geothermal energy (The Verkhovna, 2017). The highest feed-in tariff coefficients are set for electricity produced by solar power plants with an installed capacity of 10 MW or less, and the lowest for wind power plants with an installed capacity up to 600 kW. Later launch of the power plant foresees lower feed-in tariffs due to the use of lower coefficients. For example, the launch of a solar power plant (SPP) with a capacity of 10 MW or less by 31.03.2013 guaranteed the feed-in tariff rate of 8.64, while for the same capacity SES launched during the period 2025–2029 the feed-in tariff rate will be only 2.23, that is 3.87 times less. Similarly, launching a SPP in a private household in 2014 guaranteed the owners the tariff almost 2 times higher than launching of a power plant during 2017–2019 (The Verkhovna, 2017). This tariff reduction strategy aims to ensure the gradual reduction in alternative energy cost to the level of traditional one as the share of renewable energy in the overall energy balance of the country increases. It makes RE owners to use new technology that guarantee the production of electricity at the lowest cost.

The introduction of the feed-in tariff for the domestic business sector in 2009–2014 resulted in building solar and wind parks in the eastern and southern regions of Ukraine with the most favorable natural conditions for the development of these types of RES. At the same time, in 2012–2014 RES developed most actively (Table 9.1). During this period RES number increased by 44.2% with an increase in their capacities by almost 1.5 times and by 2.56 times in the amount of generated electricity. Despite such a rapid development, the share of electricity from RES in the overall power balance of Ukraine changed very slowly: from 0.56% in 2012 to 1.26% in 2014 (Sotnyk, 2017). When Ukraine lost more than a third of its green power output because of the annexation of the Crimea and a military conflict in Donbas in 2014, in 2015–2017 the country tried to compensate lost power capacity by building new RE plants. During this period alternative energy capacity increased by 1.42 times, mainly due to the growth of solar (80.1%) and bioenergy (by 48.9%) compared to 2014. The number of RE installations increased by 89.4%, while electricity generation in the sector increased by 19.3%, excessively compensating the reduction in green electricity production caused by the loss of capacities in the Crimea and the Donbas for three years.

In an effort to achieve the national goals for RE sector development, starting from 2015, the state government has expanded the circle of participants in the alternative energy market by granting the right to trade renewable

Table 9.1 The main indicators of RE sector development in Ukraine in 2012–2017 (Kurbatova, 2018; NCSREPU, 2018a)

Type of RE Power Plant*	2012	2013	2014	2015	2016	2017
Number of RE power plants in Ukraine						
SPP	41	83	70	84	126	193
WPP	14	17	13	13	16	20
SHPP	80	90	102	114	125	137
BioPP	3	9	14	16	18	27
Total	138	199	199	227	285	377
Total installed capacity of RE power plants in Ukraine, MWt						
SPP	371.6	748.4	411.9	431.7	530.9	741.9
WPP	193.8	339.1	426.1	426.1	437.8	465.1
SHPP	73.5	75.3	80.2	86.9	90	94.6
BioPP	6.2	23.7	49.1	52.4	59.1	73.1
Total	645.1	1186.5	967.3	997.1	1117.8	1374.7
Amount of generated electricity by RE power plants in Ukraine, million kWh						
SPP	334.0	562.8	485.2	464.71	487.0	714.7
WPP	257.5	636.5	1171.4	973.68	924.5	973.5
SHPP	172.0	285.9	250.6	171.6	189.3	494.8
BioPP	21.2	37.2	100.2	114.13	170.4	212.5
Total	784.7	1522.4	2007.4	1724.12	1771.2	2395.5
The share of electricity from RES in the total electricity balance of Ukraine, %						
Total	0.56	0.93	1.26	1.28	1.32	1.47

*SPP, WPP, SHPP, BioPP – solar, wind, small hydro and bioenergy power plants accordingly.

electricity of its own production to private households. Households that installed SES can sell electricity at the feed-in tariff rates since 2015; the households that installed WES can benefit from feed-in tariffs since 2017. This policy resulted in the intensive increase of RES capacity in the residential sector. So, if in the first quarter of 2015 there were only 40 households in the country with installed solar capacity of 0.3 MW, then at the end of 2015 there were 244 households with the capacity of 2.218 MW, in 2016 – 1109 households with the capacity of 16.748 MW, in 2017 – 3010 households with the capacity of 51.002 MW (NCSREPU, 2018a; SAEEESU, 2018). Thus, over three years, the number of households generating green electricity has increased more than 75 times, and the volume of installed capacities for more than 170 times (Table 9.2). The electricity produced from private households equipped with solar installations amounted to 0.41 million kWh by 2015, in 2016 – 4.246 million kWh or more than 10 times, in 2017 – 22.659 million kWh (5.3 times more compared to the previous year). In 2017 four wind

Table 9.2   The main indicators of RE development in the household sector of Ukraine in 2015–2017 (SAEEESU, 2018; SAEEESU, 2018a)

Indicator	2015 Solar Energy	2016 Solar Energy	2017 Solar Energy	2017 Wind Energy
Number of generating units, pcs	244	1109	3010	4
Installed capacity, MW	2	17	51	0.032
Volume of electricity produced, million kWh	0.410	4.246	22.659	0.001

power plants with the installed capacity of 0.032 MW added 1149 kWh to the generation of electricity from the private solar installations (Sotnyk, 2018). Overall, 88 million euro has been invested in this sector over the three years of the existence of the feed-in tariff for households (2014–2017), (SAEEESU, 2018a).

In general, in Ukraine for 2014–2017 more than $550 million have been invested in RE (NICU, 2018). However, this is not a great achievement, as according to (CMU, 2014), in order to meet the 11% renewable energy target in the country's energy balance, it is necessary to invest €12 bn in green energy by 2020 and about €20 bn in investments by 2035 year to ensure a 25% share of RES in the energy balance according to (CMU, 2017). Thus, at the beginning of 2018, only 4.58% and 2.75% of investment plans respectively were met. Rather small domestic RES sector is not a significant employer, compared to alternative energy sector in other countries of the world. However, for example, in 2017, the construction and commissioning of new RES facilities made it possible to increase employment in the sector by 320 job places. As of 01.01.2018, 2164 workers were employed in RE sector (NCSREPU, 2018a).

Thus, the dynamic development of alternative energy in Ukraine in the long run will increase energy independence level of the country, decarbonizing the national economy, creating new jobs and fostering long-lasting economic stability. However, given low energy efficiency of the industrial and housing complex, high energy prices and expensive utilities, the "energy poverty" of the Ukrainian population, the lack of investment in green innovation in the energy sector, in particular, in the construction of RES facilities in domestic households, it is necessary to further improve the state policy for RE development. It should provide economic support to all who need it for the implementation of alternative energy projects. At the same time, such policy can both help to overcome social inequalities and disproportions in the development of Ukrainian society and deepen them, therefore, policy reform requires careful management decisions.

9.4 Threats and Opportunities to Overcome the Disparities in Energy Development of Ukraine Due to Renewable Energy

RE sector development, in particular at the expense of households, can become one of the locomotives of the Ukrainian economy that will create a significant number of new jobs, reduce the import of fossil energy resources, increase the level of income and purchasing power of the population, and stimulate the development of other sectors of the national economy. At the same time, domestic alternative energy in the residential sector is characterized by both strengths that ensure the realization of existing opportunities for using RES and weaknesses that create certain threats to the development of this segment of the energy market. The SWOT analysis results presented in Table 9.3 allow examining the main problems of the sector's development and ways to solve them.

Material, technical and technological improvements in the sector, in particular electricity accumulation facilities and possibility to transform it to the required level at an affordable price should be an important task for further RES development not only in Ukraine but also in the whole world. This will help eliminate instability and unpredictability in alternative energy generation as the main RE disadvantages and will allow its free use for various economic needs, for example, for charging electric vehicles, which will contribute to the development of cleaner electric vehicles and the decentralization of electricity generation processes. The transition to alternative energy, strengthened by developing electric vehicles, will significantly reduce greenhouse gas emissions, gas pollution of urban areas, and the morbidity of the population, etc.

The restriction of the current legislation of Ukraine on the use of feed-in tariffs by private households with solar panels only on roofs and facades of buildings is one of the negative factors that influence RES development, primarily solar power generation. At the same time, the use of rotary trackers which allow a significant increase in the generation of electricity during the light day is formally restricted. If, in such circumstances, individual trackers are used for each battery, then, due to the formation of shadow zones, the location density of the solar panels decreases as well as their installed power. The use of these trackers with the simultaneous decrease in the solar panel

Table 9.3   SWOT-analysis of RE development in the household sector of Ukraine

Strengths	Weaknesses
Great unrealized RE potential	Outdated 0.4 kV power grids that need to be upgraded
Developed power grids in the country that provide 100% access to electricity for households and businesses.	Poorly developed infrastructure for domestic RE market
Development of the national RE technologies	The minimum number of instruments for economic incentives applied to households compared to business entities
Qualified specialists able to ensure the implementation of RE projects	Limitation of RES for households (solar and wind energy) supported by the national government
State support for RE projects by providing the feed-in tariff for home energy generators	
The highest feed-in tariff rates in Europe, fixed in euro	Limits on installed capacity up to 30 kW for private households RES generators
The obligation to purchase green electricity from local energy companies at feed-in tariff rates by households to ensure confidence in electricity sales and additional income for households	Limits on the location of RES generators in private households on roofs and facades of buildings, which makes the use of trackers inappropriate and reduces the amount of electricity generation
World trends in reducing RES technology and energy storage systems cost	Lack of governmental economic support for RE projects implemented by associations of co-owners of multi-apartment buildings (ACMHs)
	Lack of governmental financial support for RE projects for households
	Intermittent energy from RES due to changes in natural conditions during the year that influence payback periods of such power plants and keeps the population from investing in RE
	Lack of efficient and affordable storage systems for green electricity accumulation and storage for its further use (for charging electric vehicles)
Opportunities	Threats
Gaining energy independence of the country, solving the problem of energy resource scarcity, increasing energy efficiency and reliability of energy supply for housing and communal services	High initial investment in building new RE capacities
	Tight credits for RE projects that do not ensure payback periods
	Lack of special financing sources for feed-in tariffs in the country that lead to the increase in weighted average price of electricity

(Continued)

Table 9.3 Continued

Opportunities	Threats
Implementation of the international obligations of the country regarding the share of RES in the structure of energy consumption	The growth of the weighted average price of electricity as the share of green electricity in the energy balance grows
Improving the quality of the environment	Energy poverty of the population, that impedes the implementation of RE projects
Reliable electricity supply for remote and mountainous areas	The system of state subsidies to pay for utilities create disincentives for the population to implement RE projects
Creating a competitive electricity market, lowering prices	
Creation of new jobs in RE sector	Fears of producers of green electricity regarding the possible non-compliance by the state of its obligations to adjust feed-in tariffs, taking into account fluctuations in the euro rate
The use of green electricity as an energy source for electric vehicles and, consequently, further electric vehicles market development	High risks of economic activity caused by economic instability in the country

generation area increase SPP cost and annual electricity generation falls down. In addition, the legislation allows only 30 kW limit capacity of private household RES generators and as well as other RE sources that can be used in such power plants (solar and wind power only). Elimination of these restrictions by granting permission to locate RE devices on home land, granting feed-in tariffs to owners, increasing marginal capacity of alternative energy power plants and expanding RES types would contribute to a significant RE increase in the country.

In addition, the areas for the potentially possible location of solar panels, as the most affordable and stable resource for electricity generation in the housing and communal sector can be doubled, subject to changes in Ukrainian legislation. This would grant residents in multi-apartment buildings the same feed-in tariff preferences as private home owners. The legal lease of roof areas and the facades of buildings to install solar panels of private investors could also be the incentive for further RE development in multi-apartment buildings. Thus, residents would receive additional rent income, and the community – extra electricity and all environmental RE benefits.

The "energy poverty" of the population is a significant obstacle for the energy efficient development of the housing and communal sector of Ukraine. Expensive essential goods and services and low income of the majority of

Ukrainians necessitates state subsidies for utilities. This situation demotivates citizens to increase the level of energy efficiency of their homes and earn on alternative energy production. Monetization of subsidies and tariff revision, taking into account its economic rationale and real purchasing power of the population, will allow Ukrainians to save money that can be invested in RE development.

Currently, profitability of RE projects for households in Ukraine ranges from 10% to 20% due to the feed-in tariff, while deposits rates in euro do not exceed 2–4%. Thus, investing in alternative energy is a profitable business. At the same time, such investments are quite large and, as a rule, cannot be realized through households' savings only. In this context, it is expected to enlarge the range of economic instruments that are currently used by the government to stimulate RE development in the residential sector, not limited to the feed-in tariff only. Financial support in the form of cheap long-term loans or partial state compensation, such as "warm loans" program, tax privileges (used by business agents), etc., would allow attracting public funds for the construction of private RE generators, (Sotnyk et al., 2018).

It is necessary to improve 0.4 kV distribution networks as the energy supply decentralization processes intensify at the local level. In case alternative energy generation from households increase, 4 kV distribution networks will take the main load of its redistribution among end users. The forecasts for electricity generation are important in order to balance the electricity market.

Substantial contribution to the development of RE sector will be the further expansion of the market for services related to the construction, adjustment and maintenance of RE generators. Available market infrastructure increases customer confidence in technical assistance and support of projects. Consequently, it is necessary to provide governmental economic incentives for RE sector through tax and credit privileges, and development of energy service contracts system, etc. Currently, there are many domestic RE research that can be commercialized, and highly skilled specialists can realize the latest achievements of scientific and technological progress.

9.5 Conclusion

General economic instability in the country leads to high risks of economic activity in RE sector and hampers the development processes in the sector. In addition, taking into account hryvnia devaluation, there are certain fears of the market RE players regarding the possible non-compliance of governmental obligations to adjust the feed-in tariff, taking into account fluctuations in euro exchange rates. Consequently, economic stabilization and strengthening of hryvnia exchange rate is a key for boosting RE investments processes.

The mechanism for financing feed-in tariffs might threaten alternative energy development in the near future. Legislation (The Verkhovna Rada, 2017) does not specify a particular financing source for this economic instrument. Therefore, today the cost paid by the state to producers of green electricity is taken into account while setting the average price for electricity in the country. It means that few producers receive profits from feed-in tariffs, and all electricity consumers pay for it. This is going to be an emerging issue as the share of green electricity in the overall energy balance increases. It has already been partially solved by the government through introduction of lower coefficients of feed-in tariff for RE generators that will be put into operation in the coming years. Additionally, the trend of falling prices for alternative energy power plants facilities will keep the increase in the average electricity price due to lower electricity generation cost under control. In this regard, it is advisable to adjust feed-in tariff rates decrease to reduction of capital costs for building RE generators as a result of technology improvement, provided that profit rates of such projects are maintained within the above mentioned 10–20%. This will allow setting more fair electricity prices in the country considering the growth of alternative energy share in the energy balance. Another way to lower alternative energy prices is the exemption green electricity from VAT.

Thus, RE development in the Ukrainian household sector requires many multidirectional transformations of managerial mechanisms that refer to technical, legal, organizational, economic, financial, social and other aspects. The practical realization of the suggested ways and recommendations for adjusting state policy in the sector is time consuming and costly but will allow overcoming the disparities in the country's energy development, bringing it closer to sustainable development goals.

Acknowledgements

The publication contains the results of research carried out within the framework of research works of the Ministry of Education and Science of Ukraine "Organizational and economic mechanisms for stimulating renewable energy development in Ukraine" (No. 0117U002254) and "Innovation management of energy efficient and resource saving technologies in Ukraine" (No. 0118U003571).

References

Association of Ukrainian-Chinese Cooperation (2018). Global investments in renewable energy sources. Retrieved from http://aucc.org.ua/uk/globalni-investitsiyi-u-vidnovlyuvani-dzherela-energiyi/ (in Ukrainian).

Chernyak, O. and Fareniuk, Y. (2015). Research of global new investment in renewable energy. *Bulletin of Taras Shevchenko National University of Kyiv. Economics*, 12(177):60–68. Retrieved from http://bulletin-eco nom.univ.kiev.ua/wp-content/uploads/2016/04/177_8-1.pdf.

CMU (2014). National renewable energy action plan for the period up to 2020: decree of the Cabinet of Ministers of Ukraine (CMU) No. 902-p, 01.10.2014. Retrieved from http://zakon5.rada.gov.ua/laws/show/902-2014-%D1%80 (in Ukrainian).

CMU (2017). Energy strategy of Ukraine for the period up to 2035 "Safety, energy efficiency, competitiveness": approved by decree of the Cabinet of Ministers of Ukraine (CMU), No. 605-p, 18.08.2017. Retrieved from http://mpe.kmu.gov.ua/minugol/control/uk/doccatalog/list?currDir=503 58 (in Ukrainian).

Dhaka Tribune (2017). Bangladesh towards 100% renewable energy. Retrieved from https://www.dhakatribune.com/tribune-supplements/tri bune-climate/2017/08/12/bangladesh-towards-100-renewable-energy.

EU Energy Poverty Observatory (2018). What is energy poverty? Retrieved from https://www.energypoverty.eu/about/what-energy-poverty.

IRENA (2018). Renewable Energy and Jobs Annual Review 2018. International Renewable Energy Agency (IRENA). Retrieved from https://www.irena.org/publications/2018/May/Renewable-Energy-and-Jobs-Annual-Review-2018.

Klopov, I. (2016). The mechanisms of state support for alternative energy sources. *Problems and Prospects of Economics and Management*, 1(5):117–124. Retrieved from http://ppeu.stu.cn.ua/tmppdf/194.pdf (in Ukrainian).

Kurbatova, T. (2018). *Economic Mechanisms to Promote Renewable Energy in Ukraine*. Vienna, Premier Publishing s. r. o., 2018.

Kurbatova, T. O. (2016). Scientific principles of organizational and economic mechanism of management for renewable energy development. Ph.D. Thesis in Economic Sciences, specialty 08.00.06 (Economics of Natural Resources and Environmental Protection). Sumy State University, Sumy (in Ukrainian).

Kurbatova, T. and Khlyap, H. (2015). State and economic prospects of developing potential of nonrenewable and renewable energy resources in Ukraine. *Renewable and Sustainable Energy Reviews*, 52:217–226.

Kurbatova, T., Sotnyk, I. and Khlyap, H. (2014). Economical mechanisms for renewable energy stimulation in Ukraine. *Renewable and Sustainable Energy Reviews*, 31:486–491.

NCSREPU (2018). Letter of the National Commission for State Regulation of Energy and Public Utilities (NCSREPU) No. 3671/17.3.2/7-18 from 16.04.2018 (in Ukrainian).

NCSREPU (2018a). Report on results of the activities of the National Commission for State Regulation of Energy and Public Utilities (NCSREPU) in 2017: approved by decree of NCSREPU, No. 360, 23.03.2018. Retrieved from http://www.nerc.gov.ua/data/filearch/Catalog3/Ric hnyi_zvit_2017.pdf (in Ukrainian).

NICU (2018). Renewable energy sector Unlocking sustainable energy potential. National Investment Council of Ukraine (NICU). Retrieved from http://publications.chamber.ua/Renewable%20energy%20sector.pdf.

Riazanova, N. O. (2017). Economic mechanisms of development of refurbishable energy. *Economy and the State*, 9:58–61. Retrieved from http://www.economy.in.ua/?op=1&z=3859&i=11 (in Ukrainian).

SAEEESU (2018). Letter of the State Agency on Energy Efficiency and Energy Saving of Ukraine (SAEEESU) No. 19-01/17/31-18, 19.04.2018 (in Ukrainian).

SAEEESU (2018a). Head of the State Agency on Energy Efficiency and Energy Saving of Ukraine (SAEEESU) – 4 years in office: achievements and plans for the development of energy efficiency and clean energy. Retrieved from http://saee.gov.ua/sites/default/files/PR_EE_RE_4_yea rs_30_08_2018.pdf (in Ukrainian).

SAEEESU (2019). Technically achievable potential for energy generation from renewable energy sources and alternative fuels. The State Agency on Energy Efficiency and Energy Saving of Ukraine (SAEEESU). Retrieved from http://saee.gov.ua/uk/activity/vidnovlyuvana-enerhe tyka/potentsial (in Ukrainian).

Sotnyk, I. (2017). Modern directions of state policy improvement for development of renewable energy in Ukraine. *Economy and Region*, 4(65):5–12 (in Ukrainian).

Sotnyk, I. (2018). Organizational and economic problems and prospects for renewable energy development in private households of Ukraine. *Economic Forum*, 3:47–56 (in Ukrainian).

Sotnyk, I., Shvets, I., Momotiuk, L. and Chortok, Y. (2018). Management of renewable energy innovative development in Ukrainian households: problems of financial support. *Marketing and Management of Innovations*, 4:150–160. http://doi.org/10.21272/mmi.2018.4-14.

The Verkhovna Rada of Ukraine (2017). On electric power industry: law of Ukraine No. 575/97-BP, 16.10.1997 (updated 11.06.2017), Retrieved from http://zakon3.rada.gov.ua/laws/show/575/97-%D0%B2%D1%80. (in Ukrainian).

The Verkhovna Rada of Ukraine (2018). On electricity market: law of Ukraine No. 2019-VIII, 10.06.2018. Retrieved from http://zakon.rada.gov.ua/laws/show/2019-19. (in Ukrainian).

U.S. Energy Information Administration (2013). International Energy Outlook 2013. U.S. Energy Information Administration (EIA). Retrieved from http://www.eia.gov/forecasts/archive/ieo13/pdf/0484%282013%29.pdf.

UN Environment, Bloomberg New Energy Finance (2018). Global trends in renewable energy investment 2018. Retrieved from https://europa.eu/capacity4dev/unep/documents/global-trends-renewable-energy-investment-2018.

World Energy Council (2018). Energy efficiency indicators. Retrieved from http://www.worldenergy.org/data/efficiency-indicators/.

Zavgorodnya, S. P. (2017). Factors of energy poverty and priorities of its overcoming. Retrieved from http://www.niss.gov.ua/content/articles/files/energet_bidnist-66a29.pdf (in Ukrainian).

10

Social Responsibility for Sustainable Development Goals

Nadiya Kostyuchenko[1] and Denys Smolennikov[2,*]

[1]Department of International Economic Relations,
Sumy State University, Ukraine
[2]Department of Management, Sumy State University, Ukraine
E-mail: dos@management.sumdu.edu.ua
*Corresponding Author

The chapter is dedicated to social responsibility and its role in reaching sustainable development goals. Social responsibility of government, corporate social responsibility (CSR) as well as social responsibility of communities in reaching sustainable development goals is described by the authors. "Think globally, act locally" is the main idea of the chapter. The interrelation between sustainable development goals and the links with social responsibility is outlined in this chapter.

10.1 Introduction

The Sustainable Development Goals (SDGs) is an intergovernmental initiative introduced in the Resolution of the United Nations General Assembly "Transforming our world: the 2030 Agenda for Sustainable Development" and adopted by all United Nations Member States in 2015. That is the reviewed and enriched version of the Millennium Development Goals (MDGs) introduced in 2000 (United Nations, 2015). The Sustainable Development Goals form the global agenda, but it could only succeed if there is a collaboration between all the actors throughout the world, including government, business sector and communities. Therefore, both local and global socially responsible actions corresponding to public interests should be applied to solve sustainable development challenges and to reach the global sustainable development goals.

10.2 Literature Review

The relationship between social responsibility and the main aspects of sustainable development is analyzed by many researchers (Ebner and Baumgartner 2008; Kleine and Hauff 2009; Elmualim 2017; Xia et al. 2018). Conceptually, the implementation of social responsibility actions is necessary to achieve sustainable development and Sustainable Development Goals in particular (United Nations, 2015). However, in most cases only social aspect of sustainability is considered, while environmental and economic aspects are marginalized.

Social dimension of sustainability in CSR standards was described by Szczuka (2015). Particularly, such main issues as employee rights, workplace management, staff governance and employee practices were identified. In this respect many scholars have provided various dimensions of CSR. In this notion, Xia et al. (2018), while analyzing construction industry, concluded that social enterprise, social procurement and education are the most important dimension of CSR that can help to reach sustainable development. Similarly, Ebner and Baumgartner (2008) argued that mixing up the terms CSR and SD is not correct and described CSR as a social component of the concept of sustainable development based on a stakeholder approach. As such, SDGs can be considered as a compass for developing strategies of social responsibility. According to Pedersen (2018), the inclusion of Sustainable Development Goals into long-term business strategies, and public-private partnerships, are the key action to achieve success for business and the government. In the same time, Elmualim (2017) considers sustainable development and CSR as two main drivers for business to promote sustainability. And it is proposed to use integrated CSR and SD index to evaluate business. In the business approach, Kleine and von Hauff (2009) interpret CSR as effective "business case for sustainable development". At the same time, the authors point to the optional, philanthropic and vogue features of CSR. Business, companies will be forced to change in future: CSR reporting will not only be a tool for promoting a company, but also for comprehensive monitoring and evaluation of the effectiveness of CSR activities in the context of sustainable development and Sustainable Development Goals achievement (Scheyvens, Banks, and Hughes, 2016).

The concept of CSR itself is very dynamic. It is impossible to generalize the concept of CSR for any business, since each business has a different social, environmental and ethical influence within its responsibility

(Moon, 2007). Business want to be evaluated globally as responsible. CSR gives a lot of opportunities for business as well as challenges. And companies that ignore the opportunities and challenges can lose their licenses and competitiveness on the market (Hamann, 2003).

10.3 Sustainable Development and Sustainable Development Goals

The concept of sustainable development was introduced in 1987 and globally accepted at the United Nations Conference for Environment and Development (known as Rio Conference) in 1992. But the idea of eco-development approach was created and internationally discussed much earlier by the Club of Rome in its report "Limits to Growth" (Meadows, 1972). Sustainable development is defined as "development that meets the needs of the present without compromising the ability of future generations to meet their own needs" (WCED, 1987). There are lots of definitions of sustainable development by the moment, but this one (presented in the Brundtland Commission's report "Our Common Future") is the most popular and still the most accepted one. Sustainable development includes three pillars: environmental, economic, and social. Therefore, it's directed to environmentally secure, economically prosperous, and socially fair development of all spheres of life.

In 2000 the Millennium Development Goals were introduced as an action program for the 21st century to be achieved for the year 2015. These eight MDGs listed the policy priorities to meet the needs of the world's poorest developing countries, and they included the following:

(1) To eradicate extreme poverty and hunger;
(2) To achieve universal primary education;
(3) To promote gender equality and empower women;
(4) To reduce child mortality;
(5) To improve maternal health;
(6) To combat HIV/AIDS, malaria, and other diseases;
(7) To ensure environmental sustainability;
(8) To develop a global partnership for development (United Nations, 2000).

The action plan "Transforming our world: the 2030 Agenda for Sustainable Development", adopted by all United Nations Member States in 2015, introduced 17 Global Sustainable Development Goals to be fulfilled by 2030 (United Nations, 2015). The goals are universal for all the countries, developed and emerging countries, and provide a wider range of tasks comparing to MDGs (e.g. to include the issue of inclusiveness). Moreover, SDGs are designed to interact, being interrelated (therefore, the progress on several SDGs or all of them will have synergy impact, which is much more than achieving only one particular SDG). SDGs provide a viable model for long-term growth to create a world that is comprehensively sustainable (economically, socially and ecologically) (BSDC, 2017).

10.4 The Concept of Social Responsibility

Social responsibility is widely discussed as corporate social responsibility being defined as "a fundamental concept whereby companies integrate social and environmental concerns into their business operations and in their inter-actions with their stakeholders on a voluntary basis" (European Commission, 2001). The concept of CSR was introduced in 1970-th as a framework for understanding the social responsibility of business companies. Thus, CSR is related to social agreement between business and the society (Steiner 1972). But the concept "social responsibility" is much wider than CSR (responsibility of business), as it includes also responsibility of individuals over society as well as the role of government in providing society with public goods.

There are several approaches to interpreting the concept of CSR. Some researches (e.g. BSR, 2003; IBLF, 2015) define CSR taking the issue of ethics into account, while others exclude this aspect (European Commission, 2001; BIS, 2011). Most of the scientists don't indicate environmental issue when defining CSR, while other (Jenkins, 2004; Muller, 2006) interpret the term "corporate social responsibility" as a synonym to the term "sustainable development".

Carroll describes social responsibility as a multilevel responsibility that can be presented as a pyramid (Figure 10.1).

Nevertheless, no environmental dimension of CSR was indicated in the initial Carroll's pyramid, it could be found inside different levels of CSR. Economic responsibility, being determined by the company's business activity, allows to meet the needs of consumers and at the same time to maximize profit. In this case the idea of eco-efficiency can be investigated, meaning a win–win situation created from integrating economy and ecology through the efficient use of natural resources (Schmidheiny, 1992). Legal responsibility means compliance of business activity with laws and with society expectations fixed in legal norms; including strict adherence of environmental standards. Ethical responsibility requires business structures that are in line with society's expectations, which are not enshrined in legal documents, but

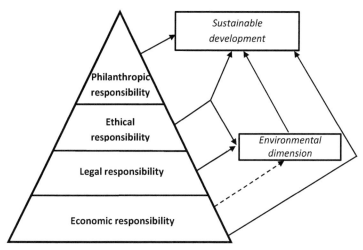

Figure 10.1 Environmental dimension of corporate social responsibility in Carroll's CSR pyramid (developed by authors based on Carroll, 1991).

are based on the norms of morality and ethics. Philanthropic responsibility prompts the business firm to take measures aimed at increasing the welfare of society through engaging in voluntary socially-oriented activities (Kolot, 2013). Both economic and legal responsibility are required by society, while ethical responsibility is expected, and philanthropic is desired (Carroll and Schwartz, 2003).

Looking deeper into the content of social responsibility (and corporate social responsibility) and knowing that sustainable development has three main dimensions, several sub-categories can be categorized, particularly: social responsibility, environmental responsibility, social and environmental charity, and finally – social and environmental responsibility (Figure 10.2). That means that interrelation of economic and social dimensions gives social responsibility (and therefore, CSR, if describing business issues). Social and environmental dimensions without economic one creates social and environmental charity. Social and environmental responsibility is a result of all three sustainable development pillars and is defined as voluntary initiatives of a company that are socially and ecologically oriented and are aimed at forming commitments to address environmental, economic and social problems in the context of sustainable development (Smolennikov and Kostyuchenko, 2015).

CSR is not a new aspect for business, since business companies are constantly experiencing the impact of external stakeholders and face not only economic but also social and environmental consequences, working

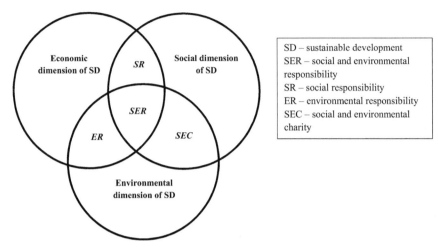

Figure 10.2   Social responsibility in frames of three dimensions of sustainable development (developed by the authors based on Carroll and Schwartz, 2003).

within the legal framework. At the same time, in the era of globalization and dissemination of the idea of sustainable development in the world, external stakeholders point to new business expectations, and national laws are changing in line with the need to take into account social, environmental and economic goals (Dahlsrud, 2008). Traditionally, CSR performed a binding function between business companies and stakeholders while the sustainability paradigm promotes actions for a fair world, providing a broader perspective for social responsibility (Kleine and von Hauff, 2009). Therefore, CSR initiatives have to be considered as fruitful strategic investments for sustainable development of companies and society (Kim et al., 2014).

10.5 Social Responsibility and Sustainable Development Goals

Sustainable development goals are a global vision for humanity, but making them real is everyone's responsibility (17 goals, 2016). Therefore, sustainable development goals can't be achieved without concrete socially-responsible everyday actions taken by all the basic stakeholders: government, communities and business.

The European Commission interprets corporate social responsibility as a corporate contribution to sustainable development (European Commission, 2001; Kleine and von Hauff, 2009). At the same time, according to the survey half of the business believe that social actions are "government territory" and responsibility (BSDC, 2017). Moreover, communities can add to reaching SDGs greatly as the global goals can be achieved only by everyday local actions of individuals (community members). As it was already mentioned there are several groups of stakeholders with different and in many cases contradictory interests that results in slowing down the process of transition towards sustainable development. At the same time, some groups of stakeholders can fasten implementation of corporate social responsibility and reaching SDGs (Smolennikov and Kostyuchenko, 2017).

10.5.1 The Role of the Government in Achieving SDGs

Government among other functions perform social function, being responsible for safety, health, honor and dignity, inviolability and well-being of citizens. Social function is enshrined in constitution and other legal documents in many countries. The government protects proper labor conditions and health of citizens, establishes the guaranteed minimum wage, provides state support for family, maternity, paternity and childhood, the disabled and the elderly, as well as provides other social guarantees. The content of social function of a government is determined by the tasks of the state in social sphere.

In 2015 all the United Nations member states officially adopted the document "Transforming our world: the 2030 Agenda for Sustainable Development" and therefore agreed to introduce SDGs into all the policies related to the key areas of economic activity, to include SDGs into the key legal documents.

The main SDGs government can operate with and fulfill social function (being responsible for those spheres) are as following:

No poverty (SDG # 1); Good health and well-being (SDG # 3);
Quality education (SGD # 4); Reduced inequalities (SDG # 10).

The government guaranties quality education, proper labor conditions and proper minimum wages in most of the countries. It puts the questions of equality into the main legal documents, trying to reduce inequalities in wages, in access to different resources (e.g. education, medical care, job positions, etc.).

The state is also responsible for clean environment. It controls quality of drinking water and sanitation (SDG # 6), operates with climate change (SDG # 13) and stimulates both companies and households to use affordable and clean energy (SDG # 7). Governments also support sustainable industrialization, clean technologies, innovations and resilient infrastructure

(SDG # 9) mostly by environmental tax system. All these, along with health care system in the country results in higher life expectancy (as a result of good health) and quality of life, including well-being (SDG # 3). Many governments have initiated reporting policies and regulation in recent years. At least 180 national policies and initiatives on sustainability reporting exist worldwide, and approximately two thirds of these are mandatory (SDG Compass, 2015).

Government is responsible for managing the quality of natural environment. The level of social and environmental safety in a region is influenced by authorities by setting up the requirements to the level of emissions and discharges of pollutants caused by industrial activity. Government can adjust business activities towards implementation of corporate social responsibility through a system of fines and restrictions as well as other economic, legal and administrative tools (Smolennikov and Kostyuchenko, 2017). However, besides mandatory legal documents, businesses and communities should make their own voluntary efforts to achieve SDGs.

10.5.2 The Role of Business Sector in Reaching SDGs

Nevertheless, the SDGs have been agreed by the governments, yet their success relies greatly on action and collaboration by all stakeholders, and business sphere has crucial role in achieving most of the sustainable development goals. Business is a vital partner in reaching SDGs as companies can contribute through their core activities (SDG Compass, 2015). Some of the SDGs correspond to the main aim of doing business, while the other sustainable development goals seem to lie beyond the profit maximization and, from the first point of view, even seem to be contradictory to that purpose. Many businessmen believe that operating with environmental and social issues adds to the production costs and give no additional benefits. At the same time, many companies used to view sustainable development as corporate social responsibility and as a chance to build their reputation and reduce waste (BSDC, 2017). Nevertheless, sustainable development is much more than CSR, companies can benefit from implementing sustainable development principles into their business processes (e.g. through getting competitive advantages by joining global reporting initiative). Global reporting initiative (GRI) helps to find out companies' impact on basic sustainability issues such as climate change, human rights, governance and social well-being (GRI, 1997).

Business being socially responsible has to apply a new sustainability strategy and clean or saving technologies to meet the sustainable development goals. According to the market study (BSDC, 2017), there is a number of specific business opportunities related to achieving the SDGs in four economic spheres, particularly:

- Food and agriculture;
- Cities;
- Energy and materials;
- Health and well-being.

These spheres were chosen because of their economic impact (generating business revenues and savings) and relevance to reaching SDGs. The fastest-growing sustainable market opportunities related to achieving SDGs are presented in Table 10.1.

Business can add greatly to reaching the SDG targets through implementing corporate social responsibility. CSR is voluntary activities of a business firm, however if there are requests from stakeholders (local communities or community members) claiming for corporate social responsibility, the reputation of the company will depend on the firm's reaction to those requests. Reputation of a company is a factor of competitive advantage in the market, therefore, on a way to sustainable development business company should consider its corporate social responsibility, including its impact on the environment (Smolennikov and Kostyuchenko, 2017). Particularly, business sphere creates working places and helps to reach sustainable development goals "No poverty" (SDG # 1) and "Zero hunger" (SDG # 2) (United Nations, 2015). This is a business sphere which can help to reduce food wastes and redistribute food surplus.

Industrial emissions influence health and therefore, well-being of human beings (that is SDG # 3). Business firms can add to SDG # 6, taking responsibility for clean water and sanitation in the region (as companies can pollute river basin and land in the area with industrial wastes, or guaranty its cleanliness through technologies and actions implemented). All this additionally influence SDG # 15 "Life on land" and SDG # 14 "Life below water" (United Nations, 2015).

When implementing CSR, companies are responsible for not polluting the environment (they implement new innovative technologies helping to reduce air pollution, water and land pollution as well as amount of materials used, and wastes produced). Additionally, industrial companies can guaranty reduction of negative impact on the environment through implementing innovations, changing infrastructure (SDG # 9), applying renewable energy resources and green technologies to produce clean energy (SDG # 7) and

Table 10.1   60 biggest market opportunities related to delivering the SDGs (BSDC, 2017)

Food and Agriculture	Cities	Energy and Materials	Health and Well-Being
Reducing food waste in values chain	Affordable housing	Circular models – automotive	Risk pooling
Forest ecosystems services	Energy efficient buildings	Expansion of renewables	Remote patient monitoring
Low-income food markets	Electric and hybrid vehicles	Circular models – appliances	Telehealth
Reducing consumer food waste	Public transport in urban areas	Circular models – electronics	Advanced genomics
Product reformulation	Car sharing	Energy efficiency – non-energy intensive industries	Activity services
Technology in large-scale farms	Road safety equipment	Energy storage systems	Detection of counterfeit drugs
Dietary switch	Autonomous vehicles	Resource recovery	Tobacco control
Sustainable aquaculture	ICE vehicle fuel efficiency	End-use steel efficiency	Weight management programs
Technology in smallholder farms	Building resilient cities	Energy efficiency – energy intensive industries	Better disease management
Micro-irrigation	Municipal water leakage	Carbon capture and storage	Electronic medical records
Restoring degraded land	Cultural tourism	Energy access	Better maternal and child health
Reducing packaging waste	Smart metering	Green chemicals	Healthcare training
Cattle intensification	Water and sanitation infrastructure	Additive manufacturing	Low-cost surgery
Urban agriculture	Office sharing	Local content in extractives	
	Timber buildings Durable and modular buildings	Shared infrastructure Mine rehabilitation Grid interconnection	

implementing responsible production, that is a part of SDG # 12 "Sustainable consumption and production" (United Nations, 2015).

Sustainable production means resource efficient and non-polluting production that results in environmentally friendly products (Ryden et al., 2013). Business companies adopting environmental policies should address not only manufacturing process, but also design, distribution, the way of use and waste processing of products (Nilsson et al., 2007). All these should be accompanied by corporate social responsibility policy to satisfy sustainable production idea. Sustainable production is essential for reaching global sustainable development goals (United Nations, 2002), as it leads to solving several other SDGs, e.g. SDG # 3, SDG # 6, SDG # 13, SDG # 14, SDG # 15. It can also help to reach such SDGs as SDG # 11, SDG # 7, SDG # 1, etc. (United Nations, 2015).

10.5.3 The Role of Communities in Achieving SDGs

Communities can include sustainable development goals into their strategies and development programs and can have a win-win situation in the long-run, benefiting from their solution to pass towards sustainability (and accounting future generations' needs and interests). That means that a successful community (from a strategic point of view) is the one that utilizes available resources rationally and minimizes negative impact on the environment.

By today there are lots of positive examples within the communities throughout the world (e.g. Ukraine, People's Republic of China, Republic of Liberia), implementing alternative energy technologies (solar, wind and composting as well as solid fuel boilers) and other small grant community projects under the EU/United Nations Development Program's projects, such as the project "A community-based reconstruction program" in Northern Liberia (Fearon, Humphreys, and Weinstein, 2009), the project "Community-based approach to local development" in Ukraine (UNDP, 2010), EU/UNDP project "HOUSES" in Ukraine (UNDP, 2018), Rugao project "Hydrogen economy pilot in China" (UNDP, 2016), the project "Market transformation of energy efficient bricks and rural buildings project" (China) (UNDP, 2008), etc.

These are the projects of community-driven development. The main aim of these projects is to stimulate sustainable social and economic development of communities by facilitating initiatives of community members to solve local level problems. Such international programs give local communities opportunity to get their projects financed, but the idea of most of the financial programs is to apply co-funding mechanisms meaning that communities need to invest their own money into the project as well, and therefore to feel responsibility for the project implementation and its lasting effect. The community members have to self-organize themselves in order to establish community organizations, to design and to implement public goods' micro projects under organizational and financial support from the international organizations and the local authorities, as well as their own small financial contribution (Petrushenko et al., 2014). Additionally, there is a positive experience of microcredit schemes for solving community problems in low-income countries. Microcrediting programs are supported by a number of international organizations (such as The International Fund for Agricultural Development (IFAD), Opportunity International, Compassion International, Oxfam, UNDP) and are widely distributed in Africa and Southeast Asia (Armendariz and Morduch, 2005; Bateman, 2010).

Every community member is responsible for making human settlements inclusive, safe and sustainable, that is a part of SDG # 11 "Sustainable cities and communities" (United Nations, 2015). An important aspect is that communities are formed from community members – individuals, who may have their own goals, that could differ from the purposes of the whole community. The sustainable development goals can be achieved if community members are socially responsible for their everyday actions which correspond to SDGs: not polluting the environment (reducing wastes, sorting household garbage), taking greenhouse gas emissions into account when making decisions on purchasing electronic equipment or other machinery (and therefore meeting SDG # 13 "Climate action"); applying alternative sources of energy in their households (and as a result, using clean and affordable energy that corresponds to SDG # 7 "Affordable and clean energy"), etc. (United Nations, 2015). As far as communities are formed of individuals, everyone is responsible for forming sustainable community and meeting the targets of every SDG mentioned. Additionally, local communities and community members have an ability to influence the business firms' behavior, being in fact, the final beneficiary of implemented corporate social responsible projects. Acting as consumers of goods and services, on one hand, they can demand certified, safe and environmentally friendly products. On another hand, local community acts as so-called "third party", that is not directly related to a business company, but can feel negative impact of its business activity on the natural environment. That's why local communities can advocate corporate social responsibility of business firms (demanding clean environment, social justice and safe working conditions as well as non-financial reporting of the company). Through arranging rallies and protests, local communities can achieve an increase in waste recycling and reduction of pollutants emissions requiring compliance with environmental standards (Smolennikov and Kostyuchenko, 2017).

Social responsibility of an individuum (and therefore, a community) can be formed through education for sustainable development that according to UNESCO (UNESCO, 2017) empowers people to change the way of thinking and work towards a sustainable future, that is a part of SDG # 4 "Quality education" (United Nations, 2015). Education for sustainable development can serve as a basis for reaching such SDGs as responsible consumption (SDG # 12), climate action (SDG # 13), clean water and sanitation (SDG # 6), sustainable communities (SDG # 11) and other goals, such as gender equality (SDG # 5), partnerships for the goals (SDG # 17) as well as other goals, because educated individuals not only form a community but can run business and become a part of the governmental establishments. Moreover, individuals may join non-governmental organizations (NGOs) which can control actions of business firms, government and communities in terms of meeting the sustainable development goals and inform the society of not fulfilling the targets.

10.6 Conclusions

Nowadays our planet faces massive environmental, social and economic challenges. To solve the problems, the Global Sustainable Development Goals were defined, and have been approved by the governments worldwide as well as action plans to achieve them by 2030. Nevertheless, the SDGs operate with global problems, they are to be solved locally by concrete everyday actions that should be taken by all the basic stakeholders: government, communities and business. Therefore, communities should change their life style for a sustainable consumption, resulting in reduction of greenhouse gas emissions and solving the problem of overconsumption. Business should change technologies for clean ones, using renewables, and develop new infrastructure resulting in sustainable production. Governments are responsible for adopting regulations, implementing environmental taxes and fines supporting sustainable solutions as well as developing policy instruments to implement them.

Thus, the sustainable development goals can be achieved through social responsibility of all the basic agents, meaning incorporating public interests into their activities.

Acknowledgements

The chapter has been supported by the Ministry of Education and Science of Ukraine under the grants No. 0117U003933, No. 0118U003569, that was financed by the state budget of Ukraine.

References

Armendariz B. and Morduch, J. (2005). *The Economics of Microfinance. Cambridge*, Mass, The MIT Press.

Bateman M. (2010). *Why Doesn't Microfinance Work? The Destructive Rise of Local Neoliberalism*. London, Zed Books.

BIS (2011). *UK Government Response to European Commission Green Paper on Corporate Social Responsibility*. Retrieved from http://asse ts.publishing.service.gov.uk/government/uploads/system/uploads/attac hment_data/file/32274/11-1097-uk-government-response-eu-corporat e-governance-framework.pdf

BSDC (2017). *Better Business Better World. The Report of the Business and Sustainable Development Commission*. Retrieved from http://report.bus inesscommission.org/uploads/BetterBiz-BetterWorld_170215_012417. pdf

BSR (2003). Issues in Corporate Social Responsibility. Retrieved from https: //www.bsr.org/en/

Carroll, A. and Schwartz, M. (2003). Corporate social responsibility: A three-domain approach. *Business Ethics Quarterly*, 13(4), 503–530.

Carroll, A. (1991). The pyramid of corporate social responsibility: Toward the moral management of organizational stakeholders. *Business Horizons*, 34(4), 39–48.

Dahlsrud, A. (2008). How corporate social responsibility is defined: An analysis of 37 definitions. *Corporate Social Responsibility and Environmental Management*, 15, 1–13.

Ebner, D. and Baumgartner, R. (2006). The relationship between sustainable development and corporate social responsibility. Retrieved from http: //www.crrconference.org/Previous_conferences/downloads/2006ebner baumgartner.pdf

Elmualim, A. (2017). CSR and sustainability in FM: Evolving practices and an integrated index. *Procedia Engineering*, 180, 1577–1584. DOI: 10.1016/j.proeng.2017.04.320

European Commission (2001). *Promoting a European Framework for Corporate Social Responsibility*. Retrieved from http://ec.europa.eu/transparency/regdoc/rep/1/2001/EN/1-2001-366-EN-1-0.pdf

Fearon, J., Humphreys, M. and Weinstein, J. (2009). Can development aid contribute to social cohesion after civil war? Evidence from a field experiment in post-conflict Liberia. *American Economic Review: Papers and Proceedings*, 99(2), 287–291. DOI: 10.1257/aer.99.2.287

GRI (1997). *About GRI*. Retrieved from http://www.globalreporting.org/information/about-gri/Pages/default.aspx

Hamann, R. (2003). Mining companies' role in sustainable development: The 'why' and 'how' of corporate social responsibility from a business perspective. *Development Southern Africa*, 20, 237–254. DOI: 10.1080/03768350302957

IBLF (2015). How to Be a Successful Entrepreneur. Retrieved from http://www.iblf.org/

Jenkins, H. (2004). Corporate social responsibility and the mining industry: conflicts and constructs. *Corporate Social Responsibility and Environmental Management*, 11(1), 23–34.

Kim, M., Kim, D. and Kim, J. (2014). CSR for sustainable development: csr beneficiary positioning and impression management motivation. *Corporate Social Responsibility and Environmental Management*, 21, 14–27. DOI: 10.1002/csr.1300

Kleine, A. and von Hauff, M. (2009). Sustainability-driven implementation of corporate social responsibility: Application of the integrative sustainability triangle. *Journal of Business Ethics*, 85, 517–533. DOI: 10.1007/s10551-009-0212-z.

Kolot, A. (2013). Corporate social responsibility: Evolution and development of theoretical views. *Economic Theory*, 4, 5–26 *(in Ukrainian)*.

Meadows, D. H., Meadows, D. L., Danders, J. and Behrens III, W. W. (1972). *The Limits to Growth*. Universe Books.

Moon, J. (2007). The contribution of corporate social responsibility to sustainable development. *Sustainable Development*, 15, 296–306. DOI: 10.1002/sd.346.

Muller, A. (2006). Global versus local CSR strategies. *European Management Journal*, 24(2–3), 189–198.

Nilsson, L., Persson, P.-O., Ryden, L., Darozhka, S. and Zaliauskiene, A. (2007). *Cleaner Production – Technologies and Tools for Resource Efficient Production*. Baltic University Press, Uppsala.

Pedersen, C. S. (2018). The UN sustainable development goals (SDGs) are a great gift to business! *Procedia CIRP*, 69, 21–24. DOI: 10.1016/j.procir.2018.01.003.

Petrushenko, Y., Kostyuchenko, N. and Danko, Y. (2014). *Conceptual Framework of Local Development Financing in UNDP Projects in Ukraine. Actual Problems of Economics*, 159(9), 257–263.

Ryden, L., Andersson, C. and Lehman, M. (2013). *Sustainable Development: a Baltic University Programme Course*. Retrieved from http://www2.b alticuniv.uu.se/index.php/introduction/home.

Scheyvens, R., Banks, G. and Hughes, E. (2016). The private sector and the SDGs: The need to move beyond 'Business as Usual'. *Sustainable Development*, 24, 371–382. DOI: 10.1002/sd.1623.

Schmidheiny, S. (1992). *Changing Course: A Global Business Perspective on Development and the Environment*. MIT Press, Cambridge.

SDG Compass (2015). *The Guide for Business Action on the SDGs*. Retrieved from http://sdgcompass.org/download-guide/

Smolennikov, D. and Kostyuchenko, N. (2015). Approaches to the definition of "environmental responsibility of business". *Bulletin of the Sumy National Agrarian University. Series: Economics and Management*, 12(66), 151–156 *(in Ukrainian)*.

Smolennikov, D. and Kostyuchenko, N. (2017). The role of stakeholders in implementing corporate social and environmental responsibility. *Business Ethics and Leadership*, 1(1), 55–62. DOI: 10.21272/bel.2017.1-07

Steiner, G. (1972). Social policies for business. *California Management Review*, Winter, 17–24.

Szczuka, M. (2015). Social dimension of sustainability in CSR standards. *Procedia Manufacturing*, 3, 4800–4807, DOI: 10.1016/j.promfg. 2015.07.587

UNDP (2008). *Market Transformation of Energy Efficient Bricks and Rural Buildings Project*. Retrieved from http://www.cn.undp.org/content/chin a/en/home/operations/projects/environment_and_energy/market-transfo rmation-of-energy-efficient-bricks-and-rural-build.html

UNDP (2010). *Community Based Approach to Local Development Project*. Retrieved from http://cba.org.ua/en/

UNDP (2016). *Hydrogen Economy Pilot in China (Rugao Project)*. Retrieved from http://www.cn.undp.org/content/china/en/home/op erations/projects/environment_and_energy/china-hydrogen-economy-p ilot-in-rugao.html

UNDP (2018). *European Union/United Nations Development Programme "Home Owners of Ukraine for Sustainable Energy Solutions (HOUSES)" Project.* Retrieved from http://www.ua.undp.org/conten t/ukraine/en/home/projects/home-owners-of-Ukraine-for-sustainable-energy-solutions.html

UNESCO (2017). *Education for Sustainable Development.* Retrieved from http://en.unesco.org/themes/education-sustainable-development

United Nations (2000). *Millennium Development Goals.* Retrieved from http://www.un.org/millenniumgoals/

United Nations (2002). *Plan of Implementation of the World Summit on Sustainable Development.* Retrieved from http://sustainabledevelop ment.un.org/milesstones/wssd

United Nations (2015). *Sustainable Development Goals.* Retrieved from http://sustainabledevelopment.un.org/sdgs

Unitied Nations (2016). Sustainable Development 17 goals, *Learn about SDGs.* Retrieved from http://17goals.org/

WCED (1987). *Our Common Future.* Oxford University Press, Oxford.

Xia, B., Olanipekun, A., Chen, Q., Xie, L. and Liu, Y. (2018). Conceptualising the state of the art of corporate social responsibility (CSR) in the construction industry and its nexus to sustainable development. *Journal of Cleaner Production*, In press. DOI: 10.1016/j.jclepro.2018.05.157.

11

The Motivational Tool-Set of Decision-Making in the Context of Sustainable Development

Anna Dyachenko[1], Svitlana Tarasenko[2,*] and Oleksandra Karintseva[1]

[1]Department of Economics and Business-Administration, Sumy State University, Ukraine
[2]Department of International Economics, Sumy State University, Ukraine
E-mail: svitlana_tarasenko@ukr.net
*Corresponding Author

11.1 Introduction

The economic approach is characterized by the fact that participants of market transactions maximize utility with a stable set of preferences and accumulate optimal amounts of information and other resources on a multitude of markets. Prices and other market instruments regulate a distribution of scarce resources in the community, thus limit desires of the bargains, and coordinate their actions. So, there are presented as informational and motivational basis of decision-making. The motivation is impulse to act which is determined certain motive or group of motives; it is the process of choosing between different possible actions that define the behaviour of the subjects focused. The system software of sustainable development supposes using of motivational toolset, enabling the consumption of environmental goods and services, and the development, implementation environmental technologies.

11.2 Motivational Tools for Making Decision in the Context of Sustainable Development

The necessity to find new mechanisms and tools for environmentally-oriented development is formulated both by the state and by scientists and researchers.

Methods to solve this problem are multidimensional, and the range of questions that are being sought answers is constantly expanding. In the 21st century, the urgency of environmentalizing production technologies and processes is actualized directly by the development of human society. So, in the work on the natural benefits of nations structured the evolution of cycles of economic development. In the 90 years the fifth cycle began, the third industrial revolution, characterized by the use in the production of ICT, the beginning of the priority use of renewable energy sources, took place. In 2015, research and development of artificial intelligence systems in Germany was marked as "The Fourth Industrial Revolution," in which producers would concentrate on the production of renewable biological resources, bio-fuels and bioenergy (Urinson, 2018).

Hargroves and Smith note that by 2020 the sixth cycle of development will begin (Hargroves and Smith, 2005). It will directly relate to renewable energy, biomic, and the active use of green technologies in the world. Such tendencies of economic development appeal to the development of approaches to organization and regulation of environmentally-oriented development of enterprises.

The urgency of the problem is also explained by the fact that environmental costs are internals of the enterprise, therefore, to stimulate business entities in ecologically oriented activities, it is necessary to apply tools that allow them to externalize their environmental costs.

Today, it is generally recognized that differences in the level of economic and social development of countries, by power of which some of them are leaders, others are lagging behind, the others – overtaken, largely depend on the institutions formed in them. And the dynamic balance in modern markets is largely determined by the institutional environment, but not just by the current changes in demand, supply and competition.

Most of the problems arising during the realization of the concept of sustainable development are also institutional, since economic entities pursue mainly the main goal of profit making and do not react to the need to ecologies production. The weakening of global ecological and economic contradictions in the absence of the possibility of radical restriction of production and consumption of goods (and as a result, reduce human impact on the environment) involves the development of a set of compensatory measures that are implemented both within a single country and with the participation of most countries. These processes are accompanied by the interaction of the processes of institutionalization at the global and national levels. For example, difficulties in penetration of biofuels into the market are related to the complexity of its codification, existing trade restrictions and quotas.

Widely used tools of institutional economics to address environmental problems began relatively recently, in the late twentieth – early twenty-first century, but today there are a number of basic operations, both domestic and foreign scientists who laid the foundation of a new methodology.

Most of the existing work, which proposes approaches to the definition of "institute", can be divided into two groups. The first group includes the work of researchers that define the institutions as rules of the game in society, provided with mechanisms for coercion to comply with these rules. The founder of this approach is North. The second group consists of the work of economists who consider institutions as balance (that is, the totality of balance strategies of the participants in the interaction), which consist of repetitive games. The basis of this approach was laid by Shotter in his work "Economic Theory of Social Institutions." The institutes were considered as balance, which solved the problems of cooperation and coordination in completely specific, repeated interactions.

If the representatives of the first approach are more focused on the analysis of developed and implemented "outside" institutions (since in this case there is a fundamental need to develop mechanisms that would ensure enforcement of these rules), the second approach is directed at the analysis of institutes that are formed "inside" systems as a result of long evolutionary processes of interaction of individual economic entities. Thus, the rules of the game are supplemented by the rules of control. And in their full definition, institutions represent a single set of rules of the game and rules of control.

Formation and change of environmental institutes take time. But it is possible to achieve a significant acceleration of the pace of change, if the knowledge use right – not only technological, but above all organizational, managerial and economic, that is, to apply the concept of growing environmental institutes associated with the implementation of targeted institutional changes. Institutes are interconnected. They exist in bonds and the desired economic effect is achieved only through interconnected institutional changes.

In complex institutional systems inevitably, there are institutional gaps – between adjacent institutes, between rules of the game and rules of control, and finally between institutional levels. Such a gap means that there is no link between adjacent rules that new and existing institutions are not complementary.

One of the reasons for institutional disruption may be the discrepancy in the degree of complexity of the new environmental institute and the general state of the institutional environment. The effectiveness of a new institution depends on how much it can reduce risks, minimize transaction costs, and ensure an efficient allocation of resources. The skills of using it and a certain degree of confidence on the part of economic agents are necessary to run a complex institute, otherwise the costs of monitoring and control aimed at preventing opportunism (when one of the agents uses this institute to the detriment of and contrary to the interests of its counterparties) will be too much high and will block the use of new institutions (North, 1997).

There are two ways in which institutional changes are directed: the improvement of existing institutional models and the cultivation of new samples. In the first case, the widespread use of business practices regulated by informal norms, when reaching the critical mass and their survival for a sufficiently long time leads to the fact that informal norms, on the one hand, are rooted, and on the other – they begin to support a layer of legal norms that they restrict. Such consolidation of the institutions can lead to a levelling of development, consolidating the institutional balance at a non-optimal point as a result of the mechanism of deterioration of the selection and institution- alization of not the best practices (Radaev et al., 2005). Therefore, let us pay attention to the second way of growing ecological institutes – the cultivation of new samples, which at the outset may not correspond to the established business practices.

There are three sources of "planting material" for cultivation: institutional inventions, import of institutes, and the recombination of existing institutional forms. Import of institutions can be carried out if additional conditions are observed concerning the content not of the cultivated rules and the way how they are cultivated. Therefore, we do not take into account this method of formation of institutes (Suharev, 2011). Accordingly, in Figures 11.1–11.2, demonstrate two basic schemes for the cultivation of environmental institutes.

Figure 11.1 shows the process of natural development of institutions. First, there are precedents for solving the problem. Then decisions are selected and begin to spread. They are supported by the forces of support, hindered by the lack of complementarity of other institutions, the forces of resistance and the barriers of the majority. The state, relying on the support forces and taking into account the arguments of the opponents, seeks to achieve a balance and reaches it in the formulation of a formal norm (law).

The law, thus, becomes a means of ennobling "wild" specimens and overcoming the majority barrier. Than following the period of adaptation to the new institution, but it usually comes alive relatively easily, since it already has a history of informal testing and is not in conflict with other institutions. Noncompliance is minimized. An important point in Figure 11.1 – is the

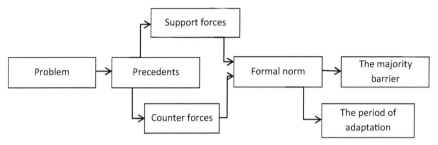

Figure 11.1 Natural development of the ecological institute.

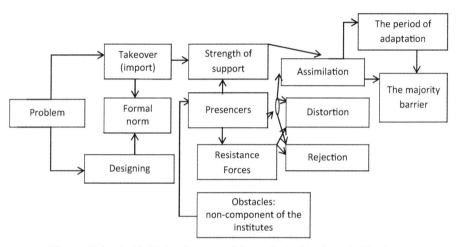

Figure 11.2 Artificial development of the ecological institute (cultivation).

formal norm appearing at the end of the process. The process of artificial planting and cultivating institutions has two options: borrowing (importing) institutes and their design, that is, design, based on internal conditions and logical constructions.

At the initial stage there is also a problem, but in the majority of cases the formal rule is formed on the basis of the best practices of other countries or is projected, based on own experience. Thus, the formal rule is at the beginning of the process, then precedents are formed – different cases of perception of the norm (assimilation, distortion or rejection). The ratio of the forces of support and counteraction is usually determined by objective obstacles that impede the assimilation of the new institution, its incomparability, inconsistency with the old institutions, different orientation in their actions. Most often motivations of support and counteraction forces are related to the desire to obtain rent, and in the process of change these forces can change places. Of course, one has to take into account whether the new transplant institution should be borrowed from the best practices or it would be advisable to take in the first stage a sample that is less conflicting with the environment.

According to Hirschmann, in the face of new formal rules, market participants adhere to three possible strategies – "voices", "loyalty" or "exit". Voice strategy is expressed in active opposition to the rule. A loyalty strategy means that market participants either actively support this rule or simply follow it in their business practices. In addition, there are always neutral forces between the forces of active opposition and support, which in many cases belong to the majority of the participants in the process. They may be neutral for two reasons: either the new rule does not affect their interests, or they choose the exit strategy. In the latter case, market participants do not protest against new orders, but simply bypass them or de-formulate new rules, implement them in existing business schemes. As a result of such "neutrality" there is a partial erosion or undermining of the formal institute. But if one manages to awaken the loyalty of neutral forces, he passes the majority barrier (Radaev et al., 2005).

Further, if the forces of support exceed the forces of counteraction and the loyal perception of the institute becomes the norm, the institute is assimilated and entered into practice. If the opposition forces prevail, then the institution at the level of informal practices is distorted or rejection and does not overcome the threshold of the majority. We have to return to the original problem, adjust the formal norm and continue to repeat the process again. The "loop" in the diagram is an illustration of an institutional trap. Another result is the rejection of the new institute, the preservation of old practices. The period of adaptation in this case is longer, sometimes the initial rejection then changes with assimilation. Therefore, it is necessary to avoid quick conclusions, to study the processes of normalization of the standards in more detail.

Along with the rejection and assimilation of the new graft-shaped formal institute, two types of intermediate states can be distinguished:

1. The emergence of parallel institutional regimes. In this case, new and old practices are being used for some time;
2. Direct distortion of the formal rule. In this case, the leading market participants are able to deform a new rule, adapting it to its narrow interests.

Thus, the internal subversion of the new institute is carried out. Formally, the new law is actively used, but it is embedded in informal business practices with results that are not in line with the planned ones. To cultivate institutions has proven to be successful, at least three elements are needed: sufficient time; special efforts; partial compromises. The time required for institutional environmental innovation to penetrate all levels of the institutional system, and rooted in them, at least to a minimum, cannot be predicted. It is necessary to provide an opportunity to show the consequences of spontaneous interactions, which often lead to adaptation, assimilation of institutions. The process of cultivating environmental institutes accordingly requires a certain amount of time. Special measures are needed to protect and support institutional innovation until it overcame the majority of the barrier. These include: administering new rules; monitoring of institutional changes; information support of innovations; learning new skills. Without developing and implementing effective mechanisms for the administration of new rules, including direct enforcement, they have little chance of proliferation. But in order for control to be effective, it is necessary to receive feedback, to regularly monitor the changes that take place. Special efforts are also needed in the field of information support related to clarifying the content of the new rules and the spread of precedents for their successful implementation in business practices. Moreover, we need not just advertising campaigns, but systematic work to promote new models, stimulating interest of leading market participants and their legitimization in the eyes of the general population. This activity should be aimed at recruiting active supporters and expanding the base of latent social support.

In the way of introducing new environmental rules, institutional compromises are needed. They mean: in order for the new institute to begin to work it is necessary to depart from its final ideal form, to form new stages of its introduction, to calculate the duration of the transition period.

Institute appears among the usual traditional institutions in response to the need to solve any problem. We regard ecological institutions as tools for solving environmental problems within the whole economy as a whole, as facilitating the transition to environmentally sustainable development can only be achieved by the state through environmental policy based on the existing macroeconomic institutional environment.

Institutional constraints should determine the direction and structure of incentives for management, reduce uncertainty and ensure the implementation of the expectations of actors. The analysis of existing institutions involves an assessment not of their form and nature of origin, but of the effectiveness. Therefore, the best environmental practices are projected to the national institutional structure and through appropriate adjustments – in the process of forming the basis of management, taking into account the continuity of endogenous (economic) methods based on exogenous (administrative) ones. It is in the context of their implementation that qualitative evolution of the institutionalization toolkit is possible – from the statement of the fact of anthropogenic influence to the formation of a culture of consumption of goods, the ecologization of social consciousness. Thus, the ecological standardization of the parameters of the functioning of the economic system is of fundamental importance and acts as a necessary component of institutionalization. However, in order to avoid declarative nature, it should be supported by monitoring and information systems that perform, including control functions.

By the sustained ecological orientation of the processes of development of economic entities, environmental institutes as a whole cover the institutional structure (this is confirmed by the processes of environmental management, excluding the possibility of substantiation and adoption of economic decisions without taking into account the environmental factor – for example, the mandatory environmental assessment of projects, environmental certification etc.).

Depending on the direction of the measures being taken, the institutions share their own environmental (for example, stimulate the development of alternative energy using renewable energy sources) and apologizing (bring the environmental component into the existing structure of public relations), such as environmental tourism, eco-labelling. The evolutionary nature of the institutional structure correction implies the simultaneous consideration of retrospective experience in solving problems and removing existing institutional disadvantages by allowing for the transplantation or borrowing of institutions that have shown effectiveness in other institutional settings and facilitates the transition to qualitatively new institutions.

Consequently, the construction of ecological and economic institutions is possible by enriching the existing institutions and cultivating new samples. In the steady ecological orientation of the processes of development of economic entities, environmental institutes as a whole cover the institutional structure. Thus, the speed of adaptation of individual economic entities, in the whole economic system to environmental change is determined by the development of environmental institutions, that is, the range of tools for regulation both formal and informal.

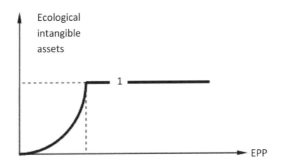

Figure 11.3 Dependence of ecological intangible assets formed on the basis of ecological competence, in the presence of one level of EPP.

In order for the entity to continuously invest in the environmentally-oriented development of the enterprise it is necessary to highlight the "ranks of ecological orientation" of enterprises.

In the case of the presence of the spectrum of positioning of enterprises on the market only on "+" and "−" (ecologically oriented enterprises and enterprises that carry out eco-design activities), the growth of the value of intangible assets formed on the basis of ecological competence of economic entities is limited (Figure 11.3). Accordingly, the company has no incentive to continue to invest in environmentally-oriented development, since it has already reached its highest level ("eco-oriented" enterprise). In the presence of, for example, three ranks of "eco-orientation," the company has additional incentives to increase eco-efficiency (Figure 11.4).

The first category of "ecological orientation" includes enterprises whose production has a minimal negative impact on the environment and (or) on human health in comparison with similar enterprises or negatively affects the environment.

The second category of "ecological orientation" includes enterprises whose production and products do not harm human health; the enterprise's production includes technologies that support reproductive processes in nature, investing in improving the quality of the environment.

The third rank of "ecological orientation" includes enterprises, whose activities meet all the criteria for sustainable development.

Thus, the presence of at least three ranks of "eco-orientation" will stimulate enterprises to continue to invest in environmentally-oriented development of the enterprise.

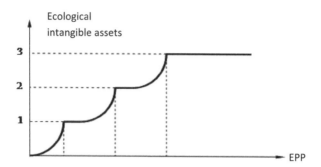

Figure 11.4   Dependence of ecological intangible assets formed on the basis of ecological competence, in the presence of three levels of EPP.

In the context of different industries and types of market structures in the enterprise, the possibility of implementing the costs of environmentalizing production processes, products, are different. Firstly, for the various industries, the amount of costs necessary for environmental production is significantly different in the volume of capital investment (for example, the cost of treatment facilities, equipment for cleaning and reuse of water for restaurant and metallurgy industries). Secondly, the effects of the use of technologies that do not harm the environment, differentiated by industry due to different returns (potential effect) of each one monetary unit of environmental costs. For example, 1 UAH of the capital investments for the installation of energy saving equipment will potentially exceed the effects of energy saving equipment in the garment production. Accordingly, the motivation to ecologies production and economic processes is also uneven. It is also determined by the possibility of including environmental costs in the price of products or services, namely, outside the competitive price. The price is determined by the market. That is, the process of externalizing environmental costs. The market reacts positively to environmental costs, the cost of goods increases, is estimated by the market – externalization is taking place. The increased cost is offset by rising prices.

For different types of market structures, the possibility of including environmental costs in the price is also different. In the market of perfect competition, the company can raise prices only if the average level of industry costs grows. That is, in the case of an individual increase in the cost of an enterprise, the price cannot be increased, and therefore environmental costs, as additional, cannot be carried over to cost. Otherwise – the loss of market shares, the un-profitable of production and economic activity (in connection with the sale at prices below cost). To create a market niche of environmental goods that differ in price from similar non-environmental goods in the market of perfect competition is quite difficult, we can talk about the lack of such an opportunity, because there are no barriers to entry, a large number of sellers, the same goods do not allow prices to be higher mid-market, and therefore include in the cost of costs associated with environmentalism.

Eco-destructive changes lead to economic losses in economic systems, as some of the economic opportunities are lost irrevocably (loss of agricultural and forest products, fixed assets, increased morbidity and mortality, low income generation), there are additional costs (for the protection of people and the prevention of diseases, re-equipment of production, construction of environmental devices and structures, the creation of corrosion-resistant materials, the use of resistant varieties of plants, etc.), as well, arose benefit is lost from the inability to operate, susceptible to disturbances of the environment.

The eco-friendliest industries in the Ukrainian economy are electric power, gas, heat and water supply (the level of ecological power, calculated on the basis of the actual loss-making estimate, on average 38%); fuel industry (26%); forestry (15%); metallurgy (12%). Taking into account expert assessments, the following list should be added: agriculture (20–25%) and transport (17–25%). The share of the environmental losses included in the official sectoral calculations accounts for just over 40% of their actual amount, calculated at the level of the economy as a whole (due to the imperfection of the existing system of generalization of statistical information in Ukraine, as well as the "shadow" Eco luminescence of non-industry sectors) (Table 11.1).

In the market of monopolistic competition products are differentiated. A sign of environmental friendliness can be considered as a basis for product differentiation. However, the deviation of the price of products from the average level may be insignificant (up to 15%), because for this market structure characterized by free access to the market, and therefore the presence of a significant number of agents that compete with each other. For example, in the market of cafes and restaurants in Sumy, the network of cafes "Strudel" is positioned in the segment "ecological food", dishes made from natural products (environmental friendliness of food). Prices vary from mid-market to 10–15% by variety of dishes.

That is, the functioning of the enterprises producing environmental goods in the market of monopolistic competition takes place within the framework of the strategy of product differentiation. The strategy of providing additional services to the main products is possible (for example, the formation and

Table 11.1 Results of the determination of the average annual level of environmental losses at the level of the Ukrainian economy (2000–2017)

		Rating Type, Million US Dollars				
Components of Environmental Losses		Costly	Actual Unprofitable	Hypothetically-Prognostic Loss-making	Lost Profits	Averaged Structure of Ecological Losses, %
1	Extraction of natural resources, incl.:	256	1214	1717	1975	13–15
1.1	Water	152	747	972	1118	8–9
1.2	Agricultural	104	467	745	857	5–6
2	Environmental pollution, incl.:	77	2172	2897	3331	22–24
2.1	Air	72	1997	2596	2985	20–22
2.2	Water	5	167	250	288	1–2
2.3	Noise pollution	–	8	9	10	<1
2.4	Electromagnetic pollution	–	–	42	48	<1
3	Waste disposal	14	104	218	251	1–2
4	Disturbance of landscapes, incl.:	–	2677	3180	3528	23–27
4.1	Erosion and soil degradation	–	1689	1801	1942	13–14
4.2	Digestion of land	–	413	619	713	5–7
4.3	Soil contamination	–	465	605	696	4–5
4.4	Land violations	–	110	155	177	1
5	The impact on biological objects	18	382	573	659	3–4
6	Emergency	2	63	317	365	1–2
7	Costs for preventing harmful effects of production	743	743	1161	1335	7–8
8	In-house environmental factors	104	111	117	135	<1–1
9	Intoxication of the population	–	1905	2540	2921	21–24
Total		1214	9371	12720	14500	100

maintenance of a fitness diet for consumers of fitness centre services that have chosen the principles of their energy conservation, recycling of waste). However, the level of competition adjusts the limits of the growth of environmental costs and the possibility of their inclusion in the price.

In the framework of the monopoly market structure, the company's ability to set the price and volume of output is formed based on the goals and the desired value of profit. Accordingly, the inclusion of the magnitude of environmental costs in the cost price is possible and is considered by the market as a positive, that is, the costs associated with the use of environmental raw materials, the installation of treatment equipment, waste recycling is externally internalized costs of the entity become external, compensated in the market.

The existence of a monopoly rent provides opportunities for variation in the behaviour and production processes of the company in terms of increasing costs for the establishment of sewage treatment facilities, the production of environmental innovations, that is, it is the monopolists and oligopolists in terms of the efficiency of economic activity can ensure the "environmental friendliness" of production processes. Also, in the monopolies and oligopoly reserves of financial resources that can be used to increase environmental safety in production, eco-friendly technology than that of economic entities operating on the market of perfect and monopolistic competition.

At the same time, the lack of competition reduces the motivation to ecologies production processes from monopolists and oligopolists. Therefore, a monopoly position serves as a means of externalizing environmental costs of the enterprise but does not ensure the functioning of the mechanism of environmentalization of the economic and industrial activities of the entity.

To stimulate the ecologization of the activities of monopoly enterprises and oligopolists, the following policies are possible:

1. Policy in the sphere of normative-legal support (regulation of environmental norms for economic entities);
2. Taxation policy (tax holidays, tax rates reduction, excise tax for enterprises implementing environmental technologies);
3. Policies in the financial and credit sector (providing credit funds at a reduced interest rate or interest-free loans for capital investments that promote the environmentalization of production processes;
4. Policies in the field of corporate social responsibility (aimed at forming corporate responsibility for the economic activity of enterprises).

At the macro level, consider motivational tools in decision making in the context of sustainable development, for example, Volkswagen.

The priority of Volkswagen is the introduction of environmental technologies. The environmental component of Volkswagen is listed in the mission.

According to the calculation of the effectiveness of Volkswagen's alternative solutions for the task of reducing the company's 25% energy consumption, water, waste, toxic substances and carbon dioxide, a comprehensive strategy is required. Based on the results of the collective work, the management offered a long-term development program that includes solving environmental problems.

The goals of Volkswagen Company were reflected in Strategy 2018. The main environmental indicators are: reduction of energy consumption by each car; CO_2 emissions (target of 95 g/km by 2020, which means consumption of 4.1 litters of gasoline or 3.6 litters of diesel fuel per 100 km); at the moment, the emissions of the vehicle make up an average of 120 g/km; investments in renewable energy sources (600 million euros by 2018), namely the development of heat and power plants, biomass fuels, as well as wind power plants; creating an infrastructure for electric vehicles, such as charging stations and utility services for repair. Such indicators may vary depending on the industry, specifics and size of the company. To increase these indicators, it is important to invest in communication and education programs on healthy lifestyles, responsible consumption, in order to raise environmental responsibility among employees, suppliers and clients (https://www.volkswagenag.com).

Volkswagen's Environmental Challenges, under the 2018 Strategy, created the Think Blue program, which involves the use of environmental technologies and efficient processes at all stages of car production and is implemented on most of the company's businesses. Within the framework of the program, work is being done on biofuels, the production of electric vehicles, increased safety of the car, control over the environmental performance of production, utilization of the car, the use of robots in the production. It should be noted that this program is implemented by a specially formed commission that controls the successful implementation of tasks at all stages of product development: from design and planning, work with suppliers, logistics, production to marketing, sales and processing.

Volkswagen research and development department continuously determines the level of harmful emissions, and the quality control department, after the decision is made by the management, implements environmental norms, standards and implements the company's environmental policy decisions. Causes of environmental deviations can be: disadvantages of cleaning

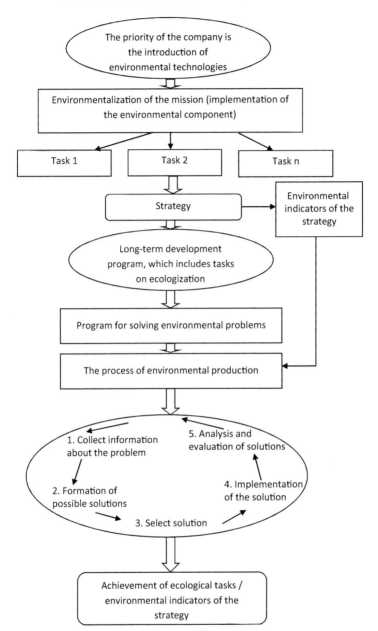

Figure 11.5 A generalized decision-making algorithm in the enterprise in the context of sustainable development.

equipment and filters; fuzzy division of responsibilities for the implementa-tion of managerial decisions; external conditions (associated with climatic conditions).

According to surveys of automotive companies, the most time-consuming stages of the algorithm of making managerial decisions in the process of environmentalizing production is the collection of information on the prob-lem of nature and implementation of the solution. This is due to the fact that at these stages involved a large number of stakeholders, engaged in monitoring the problem in the dynamics. After analysing and evaluating the results of management decisions, the decision is repeated, starting with the block identification of the problem, and in some cases, several management decisions are taken in parallel, which increases the dynamics of the passage of all blocks of the algorithm. Thus, this algorithm allows a comprehensive study of the problem and proposes an optimal solution with the maximum probability of successful implementation. The positive result of the phased implementation of the algorithm is achieved under the condition of successive use of the obtained conclusions and results that are formed at each of the previous stages.

11.3 Conclusion

The considered example of increasing the environmental performance of the automotive companies is a difficult task, but the Volkswagen algorithm has shown the possibility of implementing an approach to solving similar problems. A generalized decision-making algorithm in the enterprise in the context of sustainable development is shown in Figure 11.5.

Consequently, motivational tools in decision-making in the context of sustainable development are formed at macro and macro levels, provided by the formation of ecologically oriented institutional practices as the basis for the introduction of ecologically-oriented institutions.

References

Hargroves, K. and Smith, M. (2005) "The Natural Advantage of Nations: Business Opportunities, Innovation and Governance in the 21st Century".

North, D. (1997). Institutions, institutional changes and the functioning of the economy. Beginning, M.

Radaev, V. V., Yakovlev, A. A./Ed. and Yasina, E. G. (2005). Institutes: from bor-rowing to cultivation. The Experience of Russian Reforms and the Opportunity to Cultivate Institutional Change/Modernization of Economics and the Growth of Institutions; Higher School of Economics. – M.: Izd. House of the Higher School of Economics, 440 p.

Suharev, O. S. (2011). The Economics of the Future: the Theory of Institutional Change (evolutionary campaign) Moscow: Finance and Statistics, 121 p.

Urinson, Y. M. (2018). Industrial revolution and economic growth. – M.: Liberal Mission. 40 p. https://www.volkswagenag.com.

12

Sustainable Development Strategies in Conditions of the 4th Industrial Revolution: The EU Experience

Leonid Melnyk[1,2,*], Iryna Dehtyarova[1] and Oleksandr Kubatko[1]

[1]Department of Economics, Entrepreneurship and Business Administration, Sumy State University, Ukraine
[2]Institute for Development Economics of Ministry for Education and Science, National Academy of Science, Sumy State University, Ukraine
E-mail: melnyksumy@gmail.com
*Corresponding Author

12.1 Introduction

Success in sustainable management of human civilization development greatly depends on human ability to effectively transform economic systems towards their permanent perfection and a decrease of nature intensity of conditional production needed for human life-support. From now on, this process of economic transformation for sustainable development will be called *ecologization ("greening")* for current research. This process is a form of an integral system that stipulates permanent reproduction of basic factors of production factors including material basis, hardware and people, as well as managerial methods.

Transition of post-soviet countries to market economies necessitates the need to treat ecologization issue of public production, and to analyze all complication and variety of communications of complete production cycle and public consumption in a new way. In market systems, people's needs are the major incentive for social development in general and production in particular. In "supply – demand" framework, is a demand side that determines a long chain of supply.

In industrial technological society, that has reached climax, the productive sphere is considered to be the main point in public life. It is this sector that

241

determines current political, economic, and social processes. Human beings work with this in mind frequently forgetting that economy is only a means. Frequently, motivation of economic activity is not based on physiological needs or social interests. Very often, this damages human health, as well as, spiritual development and personal happiness. Problem of "greening" of social production is observed differently in the countries of the former-socialist block. In market systems, people's needs are the main driving force of social development and production. The "structure" in the so called "demand-consumption market structure", is considered to be a powerful engine, which propels the long chain of decisions.

The "greening" of the national economy implies a targeted process of economic transformation aimed at reduction of the integral ecological impact of the processes of production and consumption of goods and services on the environment. "Greening" is realized through a system of organized measures, innovations, restructuring of sectors of production and consumption, technological conversion, rationalization of the use of nature, and transformation of environmental protection activities at both – macro- and micro-levels.

Greening of Industry and Commerce might be considered as a function of a system which continuously reproduces interaction of the system's elements:

(1) The reproduction of green needs;
(2) The reproduction of green technological basis;
(3) The reproduction of green labor factors;
(4) The reproduction of motives for "greening" production and trade.

"Greening" of demand. Reproduction of sustainable ("green") demand is defined as permanent process of shaping the needs for sustainable goods, as well, as formation of financial possibilities for realization of identified needs. Sustainable goods are considered to be products and services that contribute to mitigation of integral ecological impact per unit of aggregate public product.

Furthermore, when speaking about reproduction of ecological needs we have to formulate the required economic conditions for "greening" of the national economy.

Firstly, the reduction of the material-energy flows of consumed goods must not lead to the lower quality of service from a standpoint of person's vital needs. Otherwise, unpredictable compensational flow of goods and services for patching up the "breaches" in consumption standards can occur. Production of these goods can lead to minimum ecological successes.

Secondly, refusal from the use of ecologically non-friendly products must be compensated by increase in the use of more ecologically friendly goods. It is necessary to meet the condition – total volume of sold goods and services (in monetary terms) as well as their production must not be reduced. It is very important because production is only one side of human activity in modern world. Even a small decrease in numerous inter-connections can lead to considerable social-economic consequences, including a decline in living conditions and an increase in unemployment. In addition, a decrease in national income can weaken the technical-scientific potential, reduce the budget of different sectors and, can worsen ecological problems. So, the demand for reproduction of the sustainable commodities is the leading link of "greening" the economy.

Finally, demand for sustainable goods must result from three interconnected economic elements: needs, elements, and possibilities. *Needs* are motives for consumption of goods that have been realized by people and communities. *Needs* are transformed into *interests*. *Demand* is undermined by financial capability, and the ability to pay for goods and services (Oosterhuis et al., 1996).

It is possible to identify four stages of development of sustainable needs.

(1) The first stage is associated with the means of control of environmental destruction ("the end of pipe").
(2) The second stage is related to environmental improvement of technology ("wasteless technology").
(3) The third stage is associated with substitution of undesirable goods and service by "greener" ones ("more efficient goods").
(4) The fourth stage is associated with production and consumption of goods for sustainable development ("sustainable life style").

Evolution of greening cycles of production and consumption of various products can be divided in two possible stages of development (Melnyk, 2006):

(1) Greening of individual components of production-consumption cycle: production, packaging, communication, storage, trade, consumption, waste disposal.
(2) Transition of economic systems from production of individual (separate) material benefits and services to formation of life-prosperity.

The "life-beneficent complex" is regarded as a destiny for human life, and as an aggregate system of material objects, cultural values, information, and natural ecosystems which ensure prosperity, as well as physical and spiritual development of people.

"Greening" of supply. Reproduction of a sustainable production is considered as generation of scientific ideas and information as well as formation of technical means which are needed for the development of green production. Social, economic, and technological causes for "greening" should and can be formulated. Social causes emerge when social interests, cultural values, and private interests of people facilitate the development of ecological needs. Economic causes are set in motion when economic conditions and organizational mechanisms make the supply of "green" goods and services profitable. Technological causes arise when there exist sufficient technical means for realization of ecological needs in production.

"Greening" people and motivational instruments. *Formation of sustainably oriented people* is considered as a continuous process of training, education and experience that provide the required information, knowledge, skills, and desire for "green" production and consumption.

"Greening" of production includes the following:

- Selection of employees with certain qualities.
- Personnel education and training.
- Ecological training and retraining.
- Development of legal standards.
- Activity regulation.
- Development of a system of rewards and penalties.
- Information.
- Control.

Motivational instruments for "greening" imply permanent facilitation of organizational, social and economic conditions, which promote the desire to achieve goals of the economy's "greening". Motivational instruments involve a system of administrative, ecological and social-physiological factors. The following are some motivational instruments in well-developed countries.

The policy and strategy of "greening". Specification of "greening" allows us to formulate local objectives for transformation of the national economy as follows:

- Restructuring of the economy.
- Restructuring of enterprises.
- Removal of needs with respect to not environmentally friendly products or services.

- Change of ecologically non-friendly technological processes.
- Lowering of the resource capacity of the products.

The principles that must be used in the process of "greening" the economy include:

- Integral approach – it stipulates the necessity to take into account all effects within the cycle of production and consumption of goods.
- Orientation according to the causes – it addresses the causes, not the consequences.
- Division of responsibility – it identifies the impact and the degree of eco-destructive activity.
- Formulation of motivational instruments under given conditions.
- Systems approach – it identifies direct and indirect influence on all objects and subjects of "greening" the economy.
- Maximum efficiency – it stipulates achievement of the goals of "greening" with minimum expenses and maximum return.

Consideration of the above-mentioned principles along with the analysis of criteria of eco-destructive influence of the production-consumption cycle allows formulation of the main directions for the "greening" of the national economy (Figure 12.1).

Evidently one can foresee the following main stages of the evolution of ecological needs. The first stage is associated with the development of the means of environmental protection from the processes of its destruction (pollution). At the second stage, priorities will be given to a substitution of ecologically non-friendly goods and services by ecologically friendly ones. There are three main strategies of "greening" the economy (Economic, 2016).

• throughput economy • wasting material • wasting energy • towards orientation products • reactive environmental protection • (over-) exploitation of natural resource • control of natural processes • throw-away mentality	• ecologically managed material flows • material productivity • energy efficiency • orientation towards function • preventive environment protection • sustainable use of natural stock • orientation towards natural processes • mentality of esteem

Figure 12.1 Conceptual directions of the development of environmental tasks (Oosterhuis, 1996).

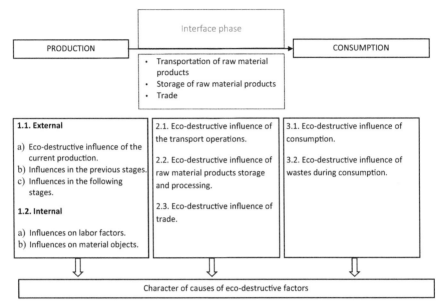

Figure 12.2   Formation of the eco-destructive factors in the main phases of the production-consumption cycle (offered by authors).

- Influence on supply, "pull-strategy" the "production-consumption train". Influencing the supply, one can pull the links of "greening" production. The essence of this strategy lies in the necessity to convince a consumer psychologically and economically to use ecologically friendly products.
- Influence on demand, called "push-strategy". The essence of this strategy is to create a system of motivational influence (ecological standards, economic instruments, information supply) which will push the producer to manufacture "green" products.
- Influence upon the communication between producers and consumers, called "interface-strategy" (as in Figure 12.2).

In "greening" the economy the intermediate links are the producers and consumers. The following are the ways of this strategy realization: the use of influential communication, mechanism for "green" trade and marketing research, and development of information systems. The use of this strategy has given an opportunity to some countries to solve very important environmental problems. The embargo on trade of rare animals and goods saved wildlife in a number of African countries. Japan managed to clean its streets and towns by introducing strict non-tariff barriers (ecological standards) on imported means of transport. Ukraine also has a list of toxic and dangerous wastes, import and transit of which are forbidden. Any country that uses these three strategies has a good chance of real success in "greening its economy". Although, a great number of economic categories have been mentioned, nonetheless basic content of different sides of the production-consumption cycle is a common factor for all these categories. Man is considered to be within this factor. Speaking about the "greening" of supply, demand, trade, and communication, means the "greening" of the relations between people. The "greening" of production and consumption can only be realized through people, labor, skills, and desires.

A phase transition to green economy. The formation of a green economy is based on a fundamentally new type of technology and economic relations. On the one hand, this is due to the need of transition to sustainable development, which allows overcoming the threat of a global environmental catastrophe and ensuring the transition to the priorities of social (personal) development. On the other hand, the achieved scientific and technical level of society at the present stage creates prerequisites for solving various corresponding problems.

Several empirical studies (Frondel and Rennings, 2007), Demirel and Kesidou (2012) stress that cost savings, reduction of resource dependence and productivity growth are key factors of eco-innovations, particularly for clean technologies. As emphasized by Frondel and Rennings (2007), innovation in clean technology tends to be driven both by cost savings, in terms of energy and material savings, environmental management systems and by regulation. The demand side determinants are mainly seen on areas with visible effect and customer benefits such as food or baby clothes. Consequently, individuals' willingness to pay a premium for organic food or organic baby clothes is substantial. Finally, also environmental process innovations create customer benefits such as less water, material or energy use.

Government intervention in green industries may be justified as a strategy to increase the supply of public goods. There are several works that tested the efficiency of different policy instruments in green industries. Thus, it is reasonable to analyze the EU countries experience of such activities.

According to Daugbjerg and Svendsen (2011) the Danish government has intervened intensively in the wind turbine industry and organic farming sector mainly for environmental reasons but with very different impact, also it should be noted that different policy instruments were used. That is to some extent such governmental interventions can be treated as a "pure experiment" within one country green industries. While the market share of wind energy reached 20 percent in 2007, organic food consumption lags behind with a market share of approximately 8.5 percent in 2007. The reason is that government intervention in the wind turbine industry has emphasized the use of policy instruments designed to increase demand for wind energy, whereas organic farming policy has put more emphasis on instruments motivating farmers to increase supply. The simple conclusion is that demand mostly represents an engine that drives innovations in green industries.

According Demirel and Kesidou (2010), the amount of resources invested into the eco-innovations depend both on internal characteristics of firm and external characteristics of environment. Thus, firms less inclined to innovations in general try to meet minimum market requirements established by consumer demand. Amount of resources invested by such firms is not big because the main purpose of production mainly profits with minimum concern to eco-innovations. Forever the stringency of economic conditions and price resources fluctuations do promote higher levels of innovations; the main explanation of such behavior is survival of the firm on the market. If the firm does not meet environmental standards it is more inclined to higher punishments and payments. More innovative firms do not necessarily need the regulatory push for eco-innovation.

Horbach (2008) used dataset based on the German Community Innovation Survey conducted in 2009 in order to test whether different types of eco-innovations are driven by different factors. It was used such explanatory variables for eco-innovations as supply, firm specific and demand factors, regulation, cost savings and customer benefits. Horbach (2008) states that fields such as material and energy savings do not need strict regulatory approaches because of their (potential) economic benefits. That is existing market motives are more than enough to stimulate such area of eco-innovations.

Considering the experience of Germany (Rave et al., 2011) econometrically proved that eco-innovators put relatively more attention to cost reduction, in particular the reduction of energy and resource costs, compared to other innovators. The last tendencies in energy price growth prevent firms from excessively using the traditionally underpriced input factor energy. Moreover, high energy costs (and their fluctuations) provide dynamic incentives to generate eco-innovations continuously. Cost factor promotes eco-innovations and contributes to the diffusion of available technologies among firm.

The formation of green economy at this stage of civilizational development becomes possible due to the fact that the Third Industrial Revolution forms prerequisites for the transition to much more efficient technological solutions for the production and consumption of goods and services.

Any phase transition is inevitably connected with the need to overcome the phase barrier caused by the immense material and social costs of the transformations that are being carried out. The transition to green economy assumes the inevitable radical transformations of society. It is precisely this task that is to be decided by the Third and Fourth Industrial Revolutions, in which human society is now rapidly entering. They are also called: Industry 3.0 and Industry 4.0.

Depending on the specific facts considered by different researchers, the emerging economy is called "green" (because it is based on environmentally friendly technologies), the "spacemen economy" (since it forms the basis for using resources in closed cycles), sustainable economy (as it focuses on achievement of sustainable development goals), post-industrial (as it is replacing the existing industrial society), information (because the leading factor is information), network (as in fact finishes the creation of a global network of local economic systems). The decisive prerequisites for the transition to a new economy are laid by events that qualitatively change the content of the three key groups of factors: material-energy, information and synergetic (communication). The following basic attributes of a new economy should appear in the nearest future:

(1) Renewable resources,
(2) Additive technologies based on 3D printers, which allow significant dematerialization of production and consumption;
(3) Distributed horizontal production networks;
(4) Solidarity and social economy components;
(5) Artificial intelligence and smart networks;
(6) Autonomous vehicles;
(7) Increase of the role of cloud technologies.

Any system is formed in the interaction of three basic groups of factors: material, information and synergetic. Their functions can be expressed in the following way: *material* – drive; *information* – direct (form the information algorithm of development); *synergetic* – unite (ensure consistent behavior of individual subsystems). Currently, synergetic factors play the leading role in this process. Their main task is to integrate separate components of local economic systems into a single systemic whole – the global economy of the Earth 'space ship'. This is exactly what happens in nature, where individual local ecosystems unite form a single biosphere of the planet.

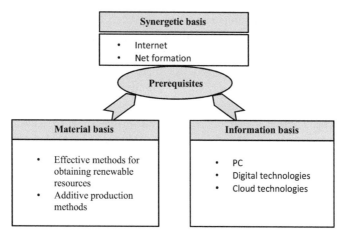

Figure 12.3   Basic prerequisites for implementation of the Third Industrial Revolution and start of the Fourth Industrial Revolution (offered by authors).

One of the most important tasks of transforming the material and energy basis of the economy during the Third Industrial Revolution is its harmonization with the natural environment. This implies, first of all, the dematerialization of the production and consumption systems, in other words, their considerable 'relief', i.e. a decrease in the material consumption and energy intensity per unit of output (work performed) and for one person living on the earth, whose vital activity must be provided with everything necessary. In addition, the task of ecological harmonization of the material and energy basis necessitates the transition to organically compatible with ecosystem metabolism substances and closed cycles of resource use.

Figure 12.3 shows the necessary basic prerequisites for the implementation of the Third Industrial Revolution. Firstly, they assume the availability of effective (i.e., cheap enough for a unit of performed work) technical means (in particular, installations of alternative energy and 3D printers); secondly, the provision of a single ('digital') basis for fixing and transmitting information (for communication between people, a man with a machine and a machine with a machine), as well as the formation of a global memory system and a kind of an all-planet 'think tank' based on cloud technologies; thirdly, the formation of a unified communication basis on the basis of the Internet and network systems.

The formation of these prerequisites created a real basis for the solution of a number of practical tasks of the sustainable transformation of the economy in conditions of the Third Industrial Revolution.

The Fourth Industrial Revolution is a logical continuation of the Third Industrial Revolution, in which a synergetic basis is the driving force of socio-economic systems development. The term "Industry 4.0" is a buzzword used widely in German speaking countries for the Fourth Industrial Revolution currently taking place. Other terms frequently used in this context are cyber-physical systems, internet of things (IoT), smart factory, smart product, big data, cloud, machine to machine (M2M) (Lang, 2016).

The Fourth Industrial Revolution concept has received great significant after the speech at the International Environment Forum in Davos (January 2016) of one of the main theorists of "Industry 4.0" phenomenon Swiss economist Klaus Schwab. He described this phenomenon as the blurring between physical, digital and biologic areas (Schwab, 2016).

For the first time the concept of the Fourth Industrial Revolution has been formulated at the Hanover Fair in 2011. The phenomenon was defined as the introduction of cyber-physical systems in production processes. Currently it is Germany that is taking the leadership in the Fourth Industrial Revolution. A public-private program 'Industry 4.0' has developed. Large German corporations having research grant support from the Federal Government are to create a fully automated production lines (smart factories), in which products interact with each other and consumers within the concept Internet of things (Khel, 2016).

New methods of teaching-learning process will allow students going deeper into the current EU issues among which social and solidarity economy and social enterprises. Social economy and social enterprises have become really very important nowadays, it is because 'they have proven to be able to engage in many and varied general interest fields of activity and tackle a variety of needs that arise in society. GECES considers it essential to highlight five positive contributions, each showing a dimension of the potential of the social economy and social enterprises to contribute to the development of the European Union' (Social, 2016).

As the example for student round table debates we may choose "The 2017 Commission Work Program". It confirms the full commitment to ensure the timely implementation of the Circular Economy Action Plan. The discussion may be around the European Commission Plastic Strategy 2017 to improve the economics, quality and uptake of plastic recycling and reuse, to reduce plastic leakage in the environment and to decouple plastics production from fossil fuels. These and many other actions will make it possible to transform socio-economic systems to sustainable development

Table 12.1   A comparison between the Neoliberal Economy and the Solidarity Economy (IDEX, 2017)

Indicators	Neoliberal Economy	Solidarity Economy
1. The role of labor	(a) Produce goods for sale (b) Provide services	(a) Satisfy needs (b) Realize your potential
2. Organization of labor	Hierarchical owners are bosses	(a) Democratic (b) Support for the group (c) Group decision-making (d) As culture
3. Technology	Substitution of human labor	An instrument for labor
4. Land	(a) Merchandise (b) Individual property (c) Business	(a) Sustenance (b) Collective property or individual property under collective use
5. Production	For the market	(a) For yourself/your family (b) To exchange (c) For the market
6. Product pricing	Depends on supply and demand	Depends on the work and its relationship to other products
7. The market	Controlled by global big business and banks	Controlled by producers and consumers, according to their real needs
8. Money	Commodity, power	A means of exchange
9. Relationships	Transactional interactions	Ongoing cooperation and building power
10. Space	Competition	Free

through green economy application. Greening the economy has to provide reduction of human footprint. Providing elements for forming green economy are:

(1) Sustainable style of life with the priority of information goods consumption;

(2) Diversification of green energy sources (solar, wind, geothermal, biogas, hydro);

(3) Deconcentrating of energy sources (hundreds of millions power units instead of hundreds ones) integrated in one EnerNet;

(4) Forming unified solidary economy on the European space.

Analyzing the above prerequisites for the achievement of sustainable development, it is possible to formulate the necessary qualities of the sessional economy, which simultaneously point to the directions on which the siding of the economy should move forward. The main ones are: resource renewability (renewable resources must become the fundamental basis of sustainable economy); dematerialization (drastic reduction in material intensity, energy intensity and environmental intensity); transformation (constant progress towards improvement through progressive transformations); innovation (perceptibility of rapid introduction of progressive innovations); naturalization (approximation of materials used, types of energy and technological processes to those that exist in nature); social orientation (the dominant goal is the transition from the priority of economic goals to the priority of social development goals); information orientation (informatization of production and consumption); ethics and humanization of economy (implementation of ethical principles of sustainable justice); synergy (the integration of individual economic entities into holistic systems ('systems of systems'), many of which acquire the scope of regional, continental or global networks); decentralization (the increased freedoms of certain economic entities in making decisions and implementing activities according to the principle: 'the center is everywhere, periphery is nowhere'); self-organization (increase the degree of systems self-organization according to the principle: 'think globally – act locally'.

Industry 3.0 is the bridge for transforming the economy to green energy systems based on renewable energy sources (RES), green technology, additive methods and 3D printers. For that reason, improved economic efficiency should provide more stimulus for poverty reducing in the world.

However, we should consider facts that new technologies can increase the inequality gap between poor and rich, since majority of resources are currently belonging to rich societies. When we speak about the environmental efficiency, the situation is reverse, and it is expected that new technologies of Third and Fourth Industrial Revolutions should help improve environmental situation. Thus, the whole clusters are successfully developing today in each of its components (solar, wind, biogas, geothermal, etc.): development of new principles for RES implementation; improvement of technological solutions (efficiency increase) within the framework of the research; efficiency increase of energy storage processes; optimization of spatial solutions for RES placement; formation of information systems optimizing RES operation and related infrastructure; the formation of communication systems that integrate the operation of certain RES into integrated energy systems. That is Third and Fourth Industrial Revolutions are mainly aimed at addressing environmental disproportions and not directly related to the reducing of intra- and international inequalities.

12.2 Conclusions

Green economy is the mainstream all over the world nowadays, a new economic model without harmful pressure on the environment. Successful implementation of renewable energy projects depends to a large extent on the efficiency of the appropriate methods of obtaining energy. In this case, the efficiency should be understood as the ratio between the assimilated amount of energy and the one that maximally reaches the source. This indicator is an analogue of efficiency coefficient used in energy industry and other technical spheres.

The efficiency of RES installations fully depends on other indicators – economic characteristics, showing the profitability of using these generators in comparison with other methods of obtaining energy. The fact is that for forty years the cost of obtaining a unit of solar energy has dropped by 150 times. In terms of cheapness it goes ahead of traditional energy. It means that the efficiency of solar energy installations is extremely increasing. New technologies and materials mean using additive technologies; here also refer new characteristics of materials as well as their environment-friendly properties. Additive technologies give unlimited design possibilities; free complexity; free provision of variability; minimum waste; production for individual needs; elimination of the assembly phase; direct materialization of information. New materials also mean materials with controlled properties; material is designed directly in the process of production.

Acknowledgements

The results of research are carried out within the framework of the projects No. 0118U003578, No. 0119U100364 and No. 0117U003935, that was financed by the state budget of Ukraine.

References

Daugbjerg, C. and Svendsen, G. T. (2011), "Government intervention in green industries: lessons from the wind turbine and the organic food industries in Denmark". Environment Development and Sustainability. vol. 13, pp. 293–307.

Demirel, P. and Kesidou, E. (2012), "On the Drivers of Eco-Innovations: Empirical Evidence from the UK". Research Policy.

European Commission (2016), Economic Commission for Europe Eighth Environment for Europe Ministerial Conference. Batumi, Georgia, 8–10 June 2016. Item 4 of the provisional agenda. Greening the economy in the pan-European region. Access mode: https://www.unece.org/fileadm in/DAM/env/greeneconomy/The_Strategic_Framework_for_GE/ece.ba tumi.conf.2016.6.e.pdf

Frondel, M. J. and Rennings, K. (2007), "End-of-Pipe or Cleaner Production? An Empirical Comparison of Environmental Innovation Decisions Across OECD Countries". Business Strategy and the Environment, vol. 16, no. 8, 571–584.

Horbach, J. (2008), "Determinants of Environmental Innovation – New Evidence from German Panel Data Sources". Research Policy vol. 37, pp. 163–173.

IDEX (2017), International Development Exchange. Available online: http s://www.idex.org/solutions/local-economies/solidarity-economy/ (Retrieved on October 19, 2018).

International debate education association. (1994), Available online: http: //idebate.org/sites/live/files/standards/documents/rules-karl-popper.pdf (Retrieved on October 19, 2018).

Khel, I. (2016), Industry 4.0: What is the Fourth Industrial Revolution? [Electronic resource]. – Access mode: http://hinews.ru/business-analitics/ind ustriya-4-0-chto-takoe-chetvertayapromyshlennaya-revolyuciya.html (Retrieved on October 19, 2018) (in Russian).

Lang Matthias. (2016), From Industry 4.0 to Energy 4.0: Future Business Models and Legal Relations, April 20, 2016. – [Electronic resource]. – Access mode: http://digitalbusiness.law/2016/04/from-industry-4-0-to -energy-4-0-future-business-models-and-legal-relations/ (Retrieved on October 19, 2018).

Melnyk, L. G. (2006), Ecological Economics. Sumy: University Book. – 367 p.

Oosterhuis, F., Rubik, F. and School, G. (1996), Product Policy in Europe: New Environmental Perspectives. Kluwer Academic Publishers. Dordrecht/Boston/London. – p. 70.

Rave, T., Goetzke, F. and Larch, M. (2011), "The Determinants of Environmental Innovations and Patenting: Germany Reconsidered". Ifo Working Paper No. 97.

Schelleman, F. (1996), From Need Assessment to Demand Management. (Discussion Document for the Greening Industry Conference, 1996, Heidelberg, Germany). The Hague. The Netherlands. – 35 p.

Schwab, K. (2016), The Fourth Industrial Revolution [Electronic resource]/K. Schwab. – Access mode: https://www.weforum.org/pages/the-fourth-industrial-revolution-by-klaus-schwab/ (Retrieved on October 19, 2018).

Social enterprises and the social economy going forward. (2016), A call for action from the Commission Expert Group on Social Entrepreneurship (GECES). October 2016. Access mode: http://ec.europa.eu/growth/tools-databases/newsroom/cf/itemdetail.cfm?item_id=9024&lang=en (Retrieved on October 19, 2018).

13

Conclusion: Inequalities with Reference to Sustainable Development Goals

Medani P. Bhandari[1,2]

[1]Akamai University, USA
[2]Finance and Entrepreneurship Department, Sumy State University, Ukraine
E-mail: medani.bhandari@gmail.com

"Inequality – the state of not being equal, especially in status, rights, and opportunities – is a concept very much at the heart of social justice theories. However, it is prone to confusion in public debate as it tends to mean different things to different people. Some distinctions are common though. Many authors distinguish "economic inequality", mostly meaning "income inequality", "monetary inequality" or, more broadly, inequality in "living conditions". Others further distinguish a rights-based, legalistic approach to inequality – inequality of rights and associated obligations (e.g. when people are not equal before the law, or when people have unequal political power)" (United Nations, 2018:1).

This book provides factual evidence of increased inequality thorough country specific case studies and comparative studies. Inequality can be seen at individuals, social groups, communities, nations; the attributes include such things as income, wealth, status, knowledge, and power (Wright, 1994). In general notion, inequality is seen every sphere of human history, socially, politically, economically and environmentally. Inequality of any nature whether it is through social, economic, cultural, political or environmental strata, creates the division on humanity and has been one of the major causes of conflict throughout human civilization.

There has been many research and publication in social sciences (mainly under the social justice paradigms and development economics paradigms), which extensively discuss various aspects of societal inequality (United Nation, 2015, 2018; Sores et al., 2014; Brockmann and Delhey, 2010;

257

Bosu et al., 2009; Adams et al., 2007; Dollar and Getti, 1999; Mander and Goldsmith, 1996; Wilson, 1996; Wright, 1994; Evans et al., 1985). The major goal of any society is sustained harmony in the society, however, it rarely exists in human sociopolitical history. The divisive element of inequality – is oppression to weak according to strata which can be economic, social, political, cultural or others forms-grounded and influenced by societal circumstances. Theoretically, it is hard to fully pinpoint the root cause of inequality – why oppressed are being oppressed and oppressors have been maintaining the domination. However, social justice theory, social dominance theory and development economics theory provides some basis of understand the nature of inequality.

What Is Social Justice?
We believe that social justice is both a process and a goal. The goal of social justice is full and equal participation of all groups in a society that is mutually shaped to meet their needs. Social justice includes a vision of society in which the distribution of resources is equitable and all members are physically and psychologically safe and secure. We envision a society in which individuals are both self-determining (able to develop their full capacities) and interdependent (capable of interacting democratically with others). Social justice involves social actors who have a sense of their own agency as well as a sense of social responsibility toward and with others, their society, and the broader world in which we live. These are conditions we wish not only for our own society but also for every society in our interdependent global community (Adam et al., 2007:1–2).

Social dominance theory is a multi – level theory of how societies maintain group – based dominance. Nearly all stable societies can be considered group – based dominance hierarchies, in which one social group – often an ethnic, religious, national, or racial one – holds disproportionate power and enjoys special privileges, and at least one other group has relatively little political power or ease in its way of life (Pratto and Stewart, 2012).

Most of the discussion about inequality, are measured through economic lenses, primarily through development economics. Development economics captures the dimensions of inequality, with the focus of well-being of the poor nations. However, socially and culturally grounded circumstances ethnicity, family background, gender, sexual orientation etc. as well as human's talent and effort part are still subjects are divisive factors which is not captured in the development paradigms.

> How inequality can be minimized?
> A capability is a potential functioning; the list of functionings is endless. It might include doings and beings such as being well nourished, having shelter and access to clean water, being mobile, being well-educated, having paid work, being safe, being respected, taking part in discussions with your peers, and so on. The difference between a capability and functioning is like one between an opportunity to achieve and the actual achievement, between potential and outcome (Walker, 2006:165).

Inequality in finding fair opportunities for the social well-beings is still matter of discussion. The discipline is still new in terms of classical economic thoughts. Effectively application of development economics can be traced after the World war II, when reconstruction of the world economic structure was in urgency.

Development Economics follows the mainstream economic principles as well considers the new variables to analyses cases of developing world, which may not capture by the main stream economic principles. This branch of economics is grounded on the basic notion of how poorest of the poor country's economic condition can be improved in holistic way covering market condition, national and international economic policies, its population and labor condition, health status, educational structure and output. There is an imbedded influence in formulating the concept of sustainable development. The notion of sustainable development deeply rooted within the frame work of development economics primarily emerged, due to different agendas of developed and developing world.

> "External trade is at best ineffective for the economic advance of less developed countries (LDCs), and more often it is damaging. Instead, the advance of LDCs depends on ample supplies of capital to provide for infrastructure, for the rapid growth of manufacturing industry, and for the modernization of their economies and societies. The capital required cannot be generated in the LDCs themselves because of the inflexible and inexorable constraints of low incomes (the various circle of poverty and stagnation), reinforced by the international demonstration effects, and by the lack of privately profitable investment opportunities in poor countries with their inherently limited local markets. General backwardness,

economic unresponsiveness, and lack of enterprise are well-high universal within the developed world. Therefore, if significant economic advance is to be achieved, governments have an indispensable as well as comprehensive role in carrying through the critical and large-scale changes necessary to break down the formidable obstacles to growth and to initiate and sustain the growth process" (Bauer 1984:27).

Above statement somehow captures the notion of development economics, by reveling the actual problem of developing world. Many theorists from classical to date have contribution in fostering development economics, however at policy formation and implementation, the United Nation has been playing important role.

13.1 Development Focuses of United Nation – Decades

- The United Nations declared the 1960s to be the decade of development. In 1961, it *"called on all member states to intensify their efforts to mobilize support for measures required to accelerate progress toward selfsustaining economic growth and social advancement in the developing countries." "At the opening of the United Nations development decade, we are beginning to understand the real aims of development and the nature of the development process. We are learning that development concerns not only man's material needs, but also the improvement of the social conditions of his life and his broad human aspirations. Development is not just economic growth, it is growth plus change."* (United Nation-Research 2019 – http://research.un.org/en/docs/dev/19 60-1970).
- In 1963, United Nations' Joint Declaration of the Developing Countries,
- 1964, UN Conference on Trade and Development (UNCTAD) establishment,
- 1965, International Cooperation Year: A/RES/1907 (XVIII), A/RES/2186 (XXI):
- 1966, Establishment of the United Nations Capital Development Fund, A/RES/2200 (XXI): International Covenant on Economic, Social and Cultural Rights, International Covenant on Civil and Political Rights and Optional Protocol to the International Covenant on Civil and Political Rights;

- 1969, Partners in development: Report of the Commission on International Development. . . .
- 1971–1980 was 2nd development decade "According to the Declaration of Mexico on the Equality of Women and their Contribution to Development and Peace, 1975 (E/CONF.66/34, para.16): "The ultimate end of development is to achieve a better quality of life for all, which means not only the development of economic and other material resources but also the physical, moral, intellectual and cultural growth of the human person." During this period, major conferences on racism, women, law of the sea, water and the environment, among others, expanded the conversation within the UN on development (United Nation-Research 2019 – http://research.un.org/en/docs/dev/1971-1980). During 2nd Development decade, United Nation began to focus on inequality issue as seen in the above statement.
- 1981–1990 was also considered as 3rd development decade. Following this train to social inclusion, in 1990, the UN Development Programme launched the first Human Development Report. *"This Report is about people – and about how development enlarges their choices. It is about more than GNP growth, more than income and wealth and more than producing commodities and accumulating capital. A person's access to income may be one of the choices, but it is not the sum total of human endeavour" (United Nation 2019).* This was another effort to empower the developing countries.
- 1991–2000 was considered human development decade. In this decade United Nation fostered and empower developing world to focus on social wellbeing of their population (health, gender, education, women empowerment).
- 2000–2015: In 2000, United Nation – declared the millennium development goals. "the Millennium Declaration identified fundamental values essential to international relations (A/RES/55/2). The Millennium Development Goals set targets for realizing these values around the world by 2015 and served as the focus for UN work throughout the period (United Nation 2019 – http://research.un.org/en/docs/dev/20 00-2015).

  Eradicate extreme poverty and hunger
  Achieve universal primary education
  Promote gender equality and empower women
  Reduce child mortality
  Improve maternal health

Combat HIV/AIDS, malaria and other diseases
Ensure environmental sustainability
Global partnership for development

- 2016–2030 – the ongoing period. The concept of development continues to evolve, as expressed in the declaration Transforming our world: the 2030 Agenda for Sustainable Development (A/RES/70/1, para. 13): *"Sustainable development recognizes that eradicating poverty in all its forms and dimensions, combatting inequality within and among countries, preserving the planet, creating sustained, inclusive and sustainable economic growth and fostering social inclusion are linked to each other and are interdependent." (United Nation 2019* – http://research.un.org/en/docs/dev/2016-2030).

    "The 2030 Agenda for Sustainable Development, adopted by all United Nations Member States in 2015, provides a shared blueprint for peace and prosperity for people and the planet, now and into the future. At its heart are the 17 Sustainable Development Goals (SDGs), which are an urgent call for action by all countries – developed and developing – in a global partnership. They recognize that ending poverty and other deprivations must go hand-in-hand with strategies that improve health and education, reduce inequality, and spur economic growth – all while tackling climate change and working to preserve our oceans and forests" (United Nations 2015).

United Nations (2015) clearly acknowledges that inequality is one of major challenge of contemporary world. As an evidence "Out of the 17 goals, eleven address forms of inequality, in terms of equality, equity and/or inclusion (Goals 1, 3, 4, 5, 6, 7, 8, 9, 10, 11, 16 and 17), and one goal (Goal 10) explicitly proposes to reduce various forms of inequalities" (Freistein and Mahlert 2015:7). This book shows how each goal are interconnected and point the difficulties part of achieving desired goals and targets of sustainable development listed in the United Nation 2015 declaration "Transforming our world: the 2030 Agenda for Sustainable Development".

In sum up

Having this general theoretical basis, this book, unveils societal inequalities through regional and country specific cases, mostly with the lense of traditional economic and development economic principles. Each chapter use and strictly follow the prescribed scholarly standards of scientific writing and provide new insights, therefore each chapter are unique.

Whereas: Chapter 1 (by Medani and Hanna) provides theoretical background/conceptual frame of the book coverage – theory and practices. Chapter 2: Economic Growth and Regional Disparities by *Shvindina Hanna, Lyeonov Serhiy, Vasilyeva Tetyana*, discuss how Inequalities, disproportions ad disparities in the process of SDGs implementation (cases). The economic and social crises and paradoxes of regional development. Territories and regions in conquest towards SDGs. Win-win strategy for the regions and territories to SDGs implementation. The principles of identification inequalities and disproportions in development.

Chapter 3: Overcoming inequalities as a source of economic development by Zharova *Liubov* provides an analysis of contemporary approaches to economic development and backgrounds of the success of developed countries. The evolution of factors of growth through changing the understanding of developing's priorities (switching from economic to social and environmental aspects). Why equality matters. The concept of "re-thinking" of modern management approaches. Generalizing of ways of overcoming common barriers of economic development.

Chapter 4: Green investments as an economic instrument to achieve SDGS by *Chygryn Olena, Pimonenko Tetyana, Lyulyov Oleksii*, show the development paradigms of green economy including mechanisms of green investments: definition, features, mechanism to implement. Comparison analyses of funding and stimulating green projects in EU and Ukraine Perspectives to provide and spread green investments in Ukraine.

Chapter 5: Reaching SDGs Through Public Institutions and Good Governance: Why Trust Matters by Buriak Anna, Oleksandr Artemenko, mostly concentrates on UN declaration "Transforming our world: the 2030 Agenda for Sustainable Development" which states "On behalf of the peoples we serve, we have adopted a historic decision on a comprehensive, far-reaching and people-centred set of universal and transformative Goals and targets. We commit ourselves to working tirelessly for the full implementation of this Agenda by 2030. We recognize that eradicating poverty in all its forms and dimensions, including extreme poverty, is the greatest global challenge and an indispensable requirement for sustainable development. We are committed to achieving sustainable development in its three dimensions – economic, social and environmental – in a balanced and integrated manner. We will also build upon the achievements of the Millennium Development Goals and seek to address their unfinished business) (United Nation 2015). The chapter analyses the essence of SDGs. The 2015 set of SDGs is considered to have an internal incompatibility caused with inherent conflict between aims of ecological and economic development. Governments are forced to elaborate adequate strategies and establish priorities in SDGs based on finance and funds they can afford. Institutions and civil society are considered to be remarkable tools for coordination of ecological and economic goals. This chapter defines the relevant factors to reach SDGs considering social conditions and economic abilities of different countries in order to increase the efficiency of expenditures through enhancing of institutions.

Likewise, Chapter 6 the concept of a sustainable economy based on labor of equal value by Zaitsev *Olexander, Shvindina Hanna, and Medani P. Bhandari*, reveals a point of view on the problem associated with "a work of equal value" and "a labor of equal value" in the contemporary world. Chapter outlines why and how civil economy is the basis of civil society including the historical account of economic development towards the civil economy. The theory of labor value is the development of the economy towards the civil economy. Indicators and measures of human labor costs. Difference and interaction of monetary and labor indicators. The concept of a sustainable economy based on labor value. How the real economy works.

Chapter 7: The mechanism of venture financing of innovative enterprises in Ukraine in the context of globalization by *I. D'yakonova, P. Podionova, O. Obod,* emphasize the venture capital as a component of finance market. The authors analyze the main trends in the development of venture market. The institutional models in a sphere of venture capital development are generalized taking into account EU experience and global tendencies. The best practices of venture companies are performed to investigate the possibilities of implementation of their patterns into Ukraine reality. The government force is analyzed as a regulator and guarantee of safety for the entities of venture financing. The strategy of sustainable development towards SDGs is modified and applied to venture financing at different levels to reduce the disproportions and inequalities.

Chapter 8: "Economic restructuring of Ukraine National Economy on the base of EU experience" by *Olena Shkarupa, Oleksandra Karintseva, Mykola Kharchenko, analyse* Ukraine National Economy in the context of SD goals with the lenses of peculiarities of scientific and methodological approach to the modernization of economic structure. The analyses providing problems and prospects of modernization of national economy according to environmental conditions as well as the role of innovations in modernization process is discussed and presented the experience of implementation of the innovations for the SD on the base of EU experience. It also focuses on the problems of costs of economic restructuring of national economy.

Chapter 9: Renewable Energy to Overcome the Disparities in Energy Development In Ukraine And Worldwide by *Sotnyk Iryna, Sotnyk Mykola, Dehtyarova Iryna, provides general outlines of* economic, social, political, and environmental benefits of renewables (RE) draw the attention of policy-makers to the issue of deploying green power capacities. Mechanisms for RE management and technology development are the research subject for many scholars. However, the practical implementation of alternative energy projects often faces organizational, economic, social and other difficulties and requires strong state support for their solution. In this regard, this chapter examines the preconditions, trends and mechanisms of RE development worldwide and in Ukraine, analyzes the issues of alternative energy sector development for domestic households as the youngest RE market participant.

Similarly, Chapter 10: Social Responsibility for Sustainable Development Goals, by *Nadiya Kostyuchenko, Denys Smolennikov* provides a theoretical evidence social responsibility and its role for sustainable development. Chapter describe social responsibility of communities as well as corporate social responsibility for reaching sustainable development goals and reveals the connection between different SDGs and the links with social responsibility. "Think globally, act locally" is the main idea we are going to stress here.

Chapter 11: The Motivational Tool set of Decision-Making in the Context of Sustainable Development by *Karintseva O.I, Dyachenko A.V., Tarasenko S.V.,* provides the theoretical and practical notion of market functions. The economic approach is characterized by the fact that participants of market transactions maximize utility with a stable set of preferences and accumulate optimal amounts of information and other resources on a multitude of markets. Prices and other market instruments regulate a distribution of scarce resources in the community, thus limit desires of the bargainers, and coordinate their actions. So, there are presented as informational and motivational basis of decision-making. The motivation is impulse to act which is determined certain motive or group of motives; it is the process of choosing between different possible actions that define the behavior of the subjects focused. The system software of sustainable development supposes using of motivational toolset, enabling the consumption of environmental goods and services, and the development, implementation environmental technologies.

Chapter 12: Sustainable development strategies in conditions of the 4th Industrial revolution: the EU experience by *Leonid Melnyk, Iryna Deht-yarova, Oleksandr Kubatko*, analyses innovative sustainable strategies in conditions of the Fourth Industrial Revolution. It reflects information, economic and technological transformations for ensuring sustainable development.

It shows the impact areas, such as production, consumption and interface where environmentally friendly decisions necessary for responding to current socio-economic and environmental challenges in the EU and Ukraine. It highlights problems, methods incentive instruments for ensuring sustainable development management and greening economy.

And finally, this chapter thirteen provides the basis of inequality in terms of social justice, dominance and development economical theoretical ground.

This book follows listed theoretical frameworks and tries to fulfil the knowledge gap of complex sustainable development discourse in reducing the global, regional, national and local inequality. The chapters result shows that even though United Nation and many other international organizations, civil society, scholars, governments have been trying to reduce the inequality-it is increased and there are no symptoms of remedy of this divisive GIANT INEQUALITY. There have been invisible walls between haves and haves not, rich and poor, developing and developed world. It is important to acknowledge, United Nation has been working vigorously since its inception to reduce the global inequality – with various declaration; however, the result has been always negative – inequality in increasing. As noted above "Out of the 17 goals, eleven address forms of inequality, in terms of equality, equity and/or inclusion (Goals 1, 3, 4, 5, 6, 7, 8, 9, 10, 11, 16 and 17), and one goal (Goal 10) explicitly proposes to reduce various forms of inequalities" (Freistein and Mahlert 2015:7). This shows the acceptance of the problems and urgency, however, there is a question "the world actually wants reduce inequality? Will sustainable development goals and targets will be achieved? Are we ready to compromise our endless desires of only me, me and mine? The invisible walls are everywhere and "my profit first" is the dominant approach of current development paradigms. To overcome the inequality problems the concept of "BashudaivaKutumbakkam" – The entire world is our home and all living beings are our relatives" and Live and let other live – the harmony within, community, nation and global" is needed.

References

Adams, M. Bell, L. A. and Griffin, P. (2007), Teaching for diversity and social justice/ – 2nd ed. Taylor & Francis Group, LLC.

Bauer, P. T. (1984), Remembrance of Studies Past: Retracing First Steps, with comments by Lipton and Srinivasan, in Meier and Seers....

Bauer, P. T. (1981), Equality, the Third World and Economic Delusion.... in Meier and Seers....

Bhandari, M. P. (2019), "BashudaivaKutumbakkam" – The entire world is our home and all living beings are our relatives. Why we need to worry about climate change, with reference to pollution problems in the major cities of India, Nepal, Bangladesh and Pakistan. Adv Agr Environ Sci. 2(1):8–35.

Bhandari, M. P. (2019), Live and let other live – the harmony with nature/living beings-in reference to sustainable development (SD) – is contemporary world's economic and social phenomena is favorable for the sustainability of the planet in reference to India, Nepal, Bangladesh, and Pakistan? Adv Agr Environ Sci. 2(2):30–60.

Bosu, R., Dare, A., Dachi, H. A. and Fertig, M. (2009), School leadership and social justice: evidence from Ghana and Tanzania. Bristol: EdQual. www.edqual.org/publications

Brockmann, H. and Delhey, J. (2010), "The Dynamics of Happiness". Social Indicators Research 97, no. 1, 387–405.

Dollar, D. and Gatti, R. (1999), Gender inequality, income, and growth: are good times good for women? World Bank Policy Research Report on Gender and Development, Working Paper Series No. 1. Washington DC: World Bank.

Evans, P., Rueschemeyer, D. and Skocpol, T. (Eds.). (1985). Bringing the state back in. Cambridge, UK: Cambridge University Press.

Freistein, K. and Mahlert, B. (2015), The Role of Inequality in the Sustainable Development Goals, Conference Paper, University of Duisburg-Essen See discussions, stats, and author profiles for this publication at: https://www.researchgate.net/publication/301675130 https://onlinelibrary.wiley.com/doi/pdf/10.1002/9780470672532.wbepp253

Mander, J. and Goldsmith, E. (1996). The case against the global economy: And for a tum toward the local. San Francisco: Sierra Club Books.

Oslington, P. and Mahmood, M. A. (1993), History of Development Economics [with Comments], The Pakistan Development Review, vol. 32, no. 4, Papers and Proceedings PART II Ninth Annual General Meeting of the Pakistan Society of Development Economists Islamabad, January 7–10, 1993 (Winter 1993), pp. 631–638 (8 pages).

Pratto, Felicia and Andrew L. Stewart, (2012), Social Dominance Theory, The Encyclopedia of Peace Psychology, First Edition. Edited by Daniel J. Christie, Blackwell Publishing Ltd.

Soares, Maria Clara Couto, Mario Scerri and Rasigan Maharajh, (eds.) (2014), Inequality and Development Challenges, Routledge, https://prd-idrc.azureedge.net/sites/default/files/openebooks/032-9/

United Nations (2018), Development Strategy and Policy Analysis Unit w Development Policy and Analysis Division Department of Economic and Social Affairs, United Nations, NY https://www.un.org/en/deve lopment/desa/policy/wess/wess_dev_issues/dsp_policy_01.pdf

Walker, M. (2006), Towards a capability-based theory of social justice for education policymaking. Journal of Education Policy, 21(2):163–185.

Walker, M., and Unterhalter, E. (Eds.). (2007), Amartya Sen's capability approach and social justice in education. Basingstoke: Palgrave Macmillan.

Wilson, W. J. (1996), When work disappears. New York: Random House.

Wright, E. O. (1994), Interrogating Inequalities. New York: Verso.

Index

About the Authors

Prof. Medani P. Bhandari completed his M.A. in Anthropology (Nepal), M.Sc. Environmental System Monitoring and Analysis (the Netherlands), M.A. Sustainable International Development (USA), M.A. and Ph.D. in Sociology (USA). He is dedicated to conservation of nature and natural resources and social empowerment through research and action project. His purpose of life is to give or contribute to the society fullest through whatever he has, earned, learned or experienced, within the Principles of "BASHUDAIVAKU-TUMBAKKAM and Live and Let Other Live". He has worked with various organizations as consultant – UNEP/Adelaide University, UNDP, IUCN, WWF, WRI, Winrock International, the Japan Environment Education Forum, and the Pajaro Jai Foundation (PJF), along with others. During 2015–17, he served as a Professor of Natural Resources and Environment at the Arabian Gulf University, Bahrain. Prof. Bhandari has spent most of his career focusing on the Sociological Theories; Environmental Sustainability; Social Inclusion, Climate Change Mitigation and Adaptation; Environmental Health Hazard; Environmental Management; Social Innovation; Developing along the way expertise in Global and International Environmental Politics, Environmental Institutions and Natural Resources Governance; Climate Change Policy and Implementation, Environmental Justice, Sustainable Development; Theory of Natural Resources Governance; Impact Evaluation of Rural Livelihood; International Organizations; Public/Social Policy; The Non-Profit Sector; Low Carbon Mechanism; Good Governance; Climate Adaptation; REDD Plus; Carbon Financing; Green Economy and Renewable Energy; Nature, Culture and Power. Prof. Bhandari's major teaching and research specialties include: Sociological Theories and Practices; Environmental Health; Social and Environmental research methods; Social and Environmental Innovation; Social and Environmental policies; Climate Change Mitigation and Adaptation; International Environmental Governance; Green Economy; Sustainability and assessment of the Economic, Social and Environmental impacts on society and nature. In brief, Prof. Bhandari has sound theoretical and practical knowledge in social science and environment science. His field experience spans across Asia, Africa, the North America,

Western Europe, Post-Soviet Countries, Australia, Japan and the Middle East. Professor Bhandari has published 4 books (his recent book is Medani P. Bhandari (2018). Green Web-II: Standards and Perspectives from the IUCN, Published, sold and distributed by: River Publishers, Denmark/the Netherlands ISBN: 978-87-70220-12-5 (Hardback) 978-87-70220-11-8 (eBook), and more than 70 scholarly papers in international scientific journals. Currently, he is serving as a Professor of Inter-Disciplinary Department – Natural Resource & Environment/Sustainability Studies, at the Akamai University, USA and Professor of the Department of Finance and Entrepreneurship, Sumy State University (SSU) Ukraine; and International Program Coordinator, Atlantic State Legal Foundation, NY, USA (remotely).

Anna Buriak graduated from Ukrainian Academy of Banking of the National Bank of Ukraine (M.S in "Banking", Ph.D. in "Money, finance and credit"), Visiting Lecturer, Technische Universität Bergakademie, Freiberg, Germany (2016), Visiting Scholar, Fachhochschule der Deutschen Bundesbank; November, "University of Applied Science", Hachenburg, Germany (2013). Currently works as Associate Professor at the Department of Finance, Banking and Insurance at Sumy State University (Ukraine). Scientific interests include Banking regulation and supervision, behavioural economics, decision-making, households, trust.

More information at http://banking.uabs.sumdu.edu.ua/index.php/en/teaching-staff/item/buriak-anna-vladimirovna

Olena Chygryn, PhD, Associate Professor, Head of Environmental Economics Section in the Department of Economics, Entrepreneurship and Business Administration at Sumy State University, Ukraine. She got the scientific degree of PhD in Economics in 2003. Chygryn Olena has published more than 100 scientific papers, including 3 papers in international peer-reviewed journals. She is the Scholarship holder of International Programs (ITEC, UNDP in Ukraine) and the participant more of 10 international training and seminars. The main sphere of her scientific interests includes: Cleaner Production, Resoursesaving, Green Entrepreneurship; Environmental Management, Corporate Governance.

More information at http://econ.fem.sumdu.edu.ua/en/list-of-employees/78-chygryn

Iryna D'yakonova graduated from Sumy Agrarian University (M.S in "Management of Organizations). Ph.D. and Habilitation in "Finance, Banking and Insurance" (Ukrainian Academy of Banking of the National Bank of Ukraine). Currently works as a Head of the Education and Research Institute for Business Technologies "UAB" and professor of the Department of the International Economic Relations at Sumy State University (Ukraine). Alumni of Applied Economic Policy Course at Joint Vienna Institute of the International Monetary Fund (2001), won ERASMUS+ KA1 scholarships (2018). Scientific interests include Supervision and regulation of Financial Markets, Venture capital as a part of finance market, International banking, Global financial crisis, Digitalization of Market Economy, Fintech in Financial Market.

More information at http://uabs.sumdu.edu.ua/en/about-the-institute/201 5-01-20-13-11-42/director

Iryna Dehtyarova graduated from Sumy State University (Specialist Diploma in Economics, Ph.D. in "Environmental Economics and Natural Resources Protection"), won international scholarships (2007–2011), and participated in international internships (2016–2017). Currently works as Associate Professor at the Department of Economics, Entrepreneurship and Business Administration at Sumy State University (Ukraine). Scientific interests include sustainable development, ecological economics, "green" economics, synergy of ecological-economic systems, socio-economic problems of the European Union. She participates in national and international scientific projects and grants financed by the EU, Flanders Government, etc.

More information at http://econ.fem.sumdu.edu.ua/en/list-of-employees /95-dehtyarova

Oleksandra Karintseva Doctor of Economic Sciences, Associate Prof of Economics, Entrepreneurship and Business Administration Department of Sumy State University (Sumy, Ukraine). Author of more than 50 scientific publications, and more than 20 papers at the scientific conferences, participated in 6 research projects and international grants (Belgium, Belarus, Ghana). Supervisor of graduates and PhD students: 6 PhD theses defended under her supervision. International activities: professional training in Germany, Belarus, Lithuania. Research interests focus on economics of sustainable development, evaluation of environmental costs, business economics, information economy, "green" economy.

More information at http://econ.fem.sumdu.edu.ua/en/list-of-employees /75-karintseva

Mykola Kharchenko Ph.D., Associate Professor of Economics, Entrepreneurship and Business Administration Department of Sumy State University (Sumy, Ukraine). The author of more than 30 scientific publications related to economics, about 10 papers at the scientific conferences, participated in 6 research projects. Participated in 2 international projects, professional training in Lithuania. Research interests: economics of enterprise, environmental economics, environmental losses, ecological and economic efficiency, ecologization, economics and management of enterprise unions activity, regional policy.

More information at https://www.researchgate.net/profile/Mykola_Kharc henko3

Nadiya Kostyuchenko graduated from Sumy State University, Ukraine (M.S. in "Environmental and Natural Resource Economics", Ph.D. in "Environmental Economics and Natural Resources Protection"), visiting research scholar at the London School of Economics and Political Science, Great Britain (2013). She won Swedish Institute scholarship (2011) and Erasmus+ scholarship (2018), was a project manager for SCOPES project "Improving Energy Security through Swiss-Ukrainian-Estonian Institutional Partnership" (2015–2017) from SNF, Switzerland.

She currently works as Associate Professor at the Department of International Economic Relations at Sumy State University (Ukraine) and as a Head of the Department for Internationalization of Educational Process. Previously she occupied a position of the Head of the Department for Teacher Training at Sumy State University. Her scientific interests are Institutional Economics, Sustainable Development, Corporate Social Responsibility, Civic Education. She actively participated in international scientific conferences and projects, including those organized by the Learning Teacher Network (UNESCO global partner), The Baltic University Program, etc. Nadiya Kostyuchenko has more than 120 academic and scientific publications. She is a member at NGO "Lifelong learning centre" and NGO "Council of young scientists".

More information at: https://www.linkedin.com/in/nadiya-kostyuchenko/

Oleksandr Kubatko is graduated from National University "Kyiv Mohyla academy" (M.S in "Economic theory"). He got the scientific degree of Dr. in Economics in 2018. Currently, Oleksandr is an Associate Professor of Economics, Entrepreneurship and Business Administration department at Sumy State University, Ukraine. He is a leader of two and a contributor of more than ten scientific and research projects (including Erasmus+, Tempus, EERC, etc.). The sphere of his scientific interests includes environmental economics, sustainable development, health economics, and EU studies.

More information at: http://econ.fem.sumdu.edu.ua/en/list-of-employees /104-kubatko

Professor Serhiy Lyeonov was graduated from the Sumy State University in 1999. He got PhD in speciality 08.02.02 – Economics and Management of Scientific and Technical Progress (2003), Associate Professor of Management Department (2006), Doctor of Economics in specialty 08.00.08 – Money, Finance and Credit (2010), Professor of Finance Department (2012). Since 2017 he is Professor at Economic Cybernetics Department. Professor Lieonov is a research leader at his department and is Business-Process Director at Sumy State University. Research interests are in a sphere of banking, investing, reproductive processes in the economy, risk management, innovation, corporate reporting and its audit, investment activity. He is the author more than 150 scientific papers and more than 50 monographs, numerous presentations at the conferences. Professor Lieonov is a Head of Specialized Academic Committee at Sumy State University, he is a mentor for many PhD students and Post-Docs. He is a contributor and a member of the Working Group of scientists who develop the forecasting of economic and social development of the Sumy region for 2018–2022 years. He is an active participant of the social projects for the local community, as well as educational events.

More information at http://cyber.uabs.sumdu.edu.ua/en/teaching-staff/it em/lieonov-serhii-en

Oleksii Lyulyov is Head of Marketing Department of Oleg Balatskyi Academic and Research Institute of Finance, Economics and Management at Sumy State University, Ukraine. He got the scientific degree of D.Sc. in Economics in 2019. Oleksii Lyulyov has published more than 120 scientific papers, including 12 papers in international peer-reviewed journals. He is the Scholarship holder of International Programs ERASMUS+ KA1 (2018) and the participant more of 10 international training and seminars. The main sphere of her scientific interests includes: country marketing policy, country image, macroeconomic stability, innovative development, sustainable economic development, strategy development, modeling and forecasting development trends.

More information at https://publons.com/researcher/1646334/oleksii-lyu lyov/

Leonid Melnyk has Doctor's Degree in Economics from Moscow Institute of National Economy; Candidate Degree of Economics from Moscow Institute of Steel and Alloys; Honors Degree of Mechanical Engineer, Kharkov Polytechnic Institute. He is a member of International Society for Ecological Economics; Honorary researcher of Xi'an University of Finance and Economics, China. Currently he works as Head of Department of Economics, Entrepreneurship and Business Administration; Director of Institute for Development Economics of Ministry for Education & Science and National Academy of Science of Ukraine at Sumy State University (Ukraine); Chairman of NGO Academy for Business and Management of Ukraine. Scientific interests include sustainable development, ecological economics, economics of systems development, economics of enterprise, comparative economics. More than 100 research and educational projects on international, national and local levels are implemented under leadership of L. Melnyk, including 30 international projects (financed by the EU, TEMPUS, TACIS, World Bank, NATO, British Council, Flanders Government, ICLEI, etc.). He is the author of more than 300 scientific publications including 30 books (monographs, textbooks, popular literature).

More information at http://econ.fem.sumdu.edu.ua/en/aboutus/employe es/head-of-department

Tetyana Pimonenko is Deputy Director for International Activity of Oleg Balatskyi Academic and Research Institute of Finance, Economics and Management at Sumy State University, Ukraine. She got the scientific degree of PhD in Economics in 2013. Tetyana Pimoneenko has published more than 90 scientific papers, including 14 papers in international peer-reviewed journals. She is the Scholarship holder of International Programs Fulbright Visiting Scholar Program (2018–2019); Latvian Government (2018); ITEC Program (India, 2018); National Scholarship Programme of the Slovak Republic for the support of mobility of students, PhD students, university teachers, researchers and artists (2017) and the participant more of 12 international training and seminars. The main sphere of her scientific interests includes: Green Marketing, Green Investment, Green Economics, Green Bonds, Alternative Energy Resources; Environmental Management and Audit in Corporate Sector of Economy, Sustainable Development and Education.

More information at http://econ.fem.sumdu.edu.ua/en/list-of-employees /318-pimonenko

Olena Shkarupa Administration Department of Sumy State University (Sumy, Ukraine). Participation in more than 30 conferences, 5 international projects (Belgium, Belarus, Lithuania), 7 scientific projects were supported by state budget, 3 international summer schools in excellence in teaching (HESP projects). Professor Shkarupa is a winner of the grant of the President of Ukraine. She is an active participant of the regional grant for analyses of the monitoring system for sustainable development in Sumy region. Research interests focus on Sustainable Development, Environmental Economics, Innovative Development, Business Planning.

More information at http://econ.fem.sumdu.edu.ua/en/list-of-employees /89-shkarupa

Shvindina Hanna graduated from Sumy State University (M.S in "Management of Organizations, Ph.D. in "Environmental Economics and Natural Resources Protection"), Post-Doc researcher (University of Montpellier, Purdue University) in "Business Studies and Management", Fulbright Alumni (2018–2019), won ERASMUS MUNDUS scholarships (2014, 2015). Currently works as Associate Professor at the Department of Management at Sumy State University (Ukraine). She is member of CENA community and Researchers' excellence network (RENET). She is member of editorial boards of several scientific journals (Ukraine, Poland, Switzerland), invited reviewer for the international conferences (Poland, Germany, Ukraine). She is author of more than 90 papers. Scientific interests include Coopetition Paradox, Strategies and Innovations, Change Management, Organizational Development, Civic Education and Leadership. She is activist and is a CEO at NGO "Lifelong learning centre", where provide courses on Leadership and Communications, organize the program on empowerment for local community.

More information at https://www.linkedin.com/in/hanna-shvindina-aa39 0924/

Denys Smolennikov is Associate Professor at the Department of Management and Deputy Director of Oleg Balatskyi Academic and Research Institute of Finance, Economics and Management at Sumy State University. In 2017

he defended the PhD thesis in "Environmental Economics and Natural Resources Protection" on the theme "Organizational and economic support of social and environmental responsibility of thermal power plants". Denys Smolennikov took an active part in a bunch of conferences and international projects devoted to the issues of sustainable development, including the conferences of The Learning Teacher Network (UNESCO global partner) in Tallinn (Estonia) and Aarhus (Denmark), the Baltic University Program in Rogow (Poland), Kaunas (Lithuania), Riga (Latvia) etc. Denys Smolennikov is the organizer of numerous national and international scientific and practical conferences and forums, in particular ISCS Economics for Ecology, STABI-CON systems, Ecological Management in the General Management System, and Economic Problems of Sustainable Development. Also, he is an active member at NGO "Lifelong learning centre" and NGO "Council of young scientists". Denys Smolennikov has around 100 academic and scientific publications, including more than 20 peer-reviewed articles in international scientific journals. Research results are implemented into the educational process, particularly included into the Teacher professional development programme on active learning techniques for education for sustainable development.

More information at http://www.linkedin.com/in/smolennikov/

Iryna Sotnyk graduated from Sumy State University (specialist in "Economics of Enterprise", Ph.D., D.Sc. in "Environmental Economics and Natural Resources Protection"), Fulbright Alumni (2019–2020), won the Cabinet of Ministers of Ukraine Prize for youth's contribution to the development of the state (2004) and scholarships of the Cabinet of Ministers (2011–2012) and the Verkhovna Rada of Ukraine (2012–2013). Currently works as Professor at the Department of Economics, Entrepreneurship and Business Administration at Sumy State University (Ukraine). Scientific interests include Economics of Energy and Resource Saving, Energy Economics, Social and Solidarity Economics. She is a Deputy Editor-in-Chief of the International Scientific Journal "Mechanism of Economic Regulation" (Sumy, Ukraine), Guest Editor of the International Scientific Journal "Global Energy Issues", a leader and a contributor of more than 25 research projects, including international ones. Under her leadership, 7 research projects including 4 grants of the President of Ukraine (2006, 2008, 2012 and 2016) were performed. Iryna Sotnyk had international training in Russia (2013), Estonia (2015), Lithuania (2016), Israel (2016).

More information at http://econ.fem.sumdu.edu.ua/en/list-of-employees /98-sotnyk

Mykola Sotnyk graduated from Kharkov Polytechnic Institute (Sumy branch) (mechanical engineer, specialty "Technology of mechanical engineering", Ph.D., D.Sc. (technical sciences)).

Currently works as Associate Professor at the Department of Applied Hydroaeromechanics at Sumy State University (Ukraine). Scientific interests include Energy and Resource Saving, Machine Building, Pump Engineering, Ecology. He is a leader and a contributor of more than 15 research and applied projects, Performing Duties Director of the Research Institute of Energy Efficient Technologies at Sumy State University. Mykola Sotnyk has great experience in conducting energy surveys and energy audits of residential and industrial buildings, technological processes in industry, housing and communal services, as well as experience in energy and hydraulic analysis of the functioning of hydraulic water supply systems, circulating water supply, drainage, modernization of electromechanical complexes operating in these systems. Has about 20 patents.

More information at http://pgm.sumdu.edu.ua/index.php/uk/hidden-men u/20-sotnik-mikola-ivanovich

Svitlana Tarasenko M.S. in Economy of Enterprise, Sumy State University Current position. C.Sc. (Economics), Senior Lecturer, Department of International Economics Relations, manager of projects at NGO "Lifelong learning centre", trainer's Program "Women's Health Ukraine" Program of the US Agency for International Development (USAID). Implementation of the project "Forming of responsible attitude towards women reproductive health "Save Life", USAID fund. Duties – support of information campaign on educational activities concerning reproductive health of women. Implementation of the project "Family planning as men's business", financed by USAID. Duties – coordinator of the project. Implementation of the projects by DVV International: "School of Civil Urbanistics: case in Sumy", "Civic Cinema: creativity as a civil action", "What you need to know for Your voice to be heard: information and financial literacy as prerequisites for the development of local democracy". Duties – coordinator or implementer of the project. Scientific interests: industrial policy, business processes, intangible assets, health care economy, quality system, sustainable development.

More information at http://kme.uabs.sumdu.edu.ua/ua/sklad-kafedri/ite m/tarasenko-svitlana-viktorivna

Professor Tetyana Vasilyeva is a Director of Oleg Balatskyi Academic and Research Institute of Finance, Economics and Management, Professor

(2010, Professor's degree of the Department of Management) and Doctor of Economics (2008, speciality 08.00.08 – money, finance and credit). She is the author of more than 200 scientific papers, 70 monographs, and numerous materials of the conferences. She is Editor-in-chief of the scientific journal "Business Ethics and Leadership", member of the editorial board of the journals: "Strategy and Development Review", "Carbon Accounting and Business Innovation Journal". She is a mentor for PhD students and Post-Docs and is a member of the Specialized Academic Committee at Sumy State University; one of the founders and a member of NGO "Council of Young Scientists". She was the initiator of launching the Business Support Center in Sumy region, and now she is a member of this organization that gained the support of the European Bank for Reconstruction and Development. She is a team leader, an author and participant of numerous scientific projects that gained the highest scores in Ukraine. Research interests include banking, investment, risk management, innovation activity, social responsibility, education. She is a leader and contributor in international projects, such as Jean Monnet projects (2014, 2015, 2017), grant projects such as "Development of Dialogue Between Banks and Civil Society in the Context of Ensuring Democratic Processes in Ukraine" (DAAD, 2016), "Enhancing Energy Security by Swiss-Ukrainian-Estonian Institutional Partnership" SCOPES IZ74Z0_160564 (Swiss National Science Foundation, 2015–2017) and many others. She is a contributor in many educational projects financed by Ministry of Economic Cooperation and Development of Germany, DVV International, DESPRO in Switzerland, American Council for Economic Education NCEE, International Foundation "Vidrodzennia" etc. Professor Vasilyeva initiated many scientific and educational events concerning the development of corporate social responsibility (e.g., winter/summer schools for youth "Economic and business values", "School of innovations and social entrepreneurship", "Management of socially responsible business", etc.).

More information at https://fem.sumdu.edu.ua/en/home-page-3-en/direk torat-gg/director

Oleksandr Zaitsev graduated from Sumy State University (1980). After graduation, he was invited to work at Sumy State University at the Department of Economics (1982). At the same time, he studied at postgraduate school of the Moscow Institute of National Economy named after G.V. Plekhanov (1989–1991). Got Ph.D. degree in "Economics of environment" in 1991.

His major is the evaluation of economic losses from air pollution, from pollution of open water resources, as well as the economic justification for the use of hydrogen fuel in engines for trucks.

Faculty member at the Department of Finance Sumy State University (1993) and Associate Professor of Finance (1996). The course provided for bachelors, masters and graduate students are in a sphere of finance. He is author of more than 50 research papers. Research interests include justification and measurement of labor (public) value indicators and their application in economic and financial assessments; questions of theory and practice of finance and taxation; financial and economic assessments of the impact of anthropogenic activities on the environment.

More information at http://fin.fem.sumdu.edu.ua/en/dep/teaching-staff/407-zaitsev-eng.html

Liubov Zharova graduated from Kyiv Polytechnic Institute with honors (M.S. "Management of organization"), obtained a Ph.D. in "Productive Forces and Regional Economics" (2005) and defended doctoral research on "Macroeconomic regulation of environmental policy: theory, methodology, practice" (2013). She permanently interacts as an expert with State Fund for Fundamental Research of Ukraine, Ministry of Ecology and Natural Resources of Ukraine, Ministry of Economic Development and Trade of Ukraine, European Bank for Reconstruction and Development and UNDP in Ukraine. Scientific interests include sustainable development and institutional support, strategic environmental assessment, equality, and contemporary management re-thinking. Currently, working as head of international economic relations, business, and management department of Ukrainian American Concordia University (Ukraine) and professor University of Economics and Humanities (Poland). She is an activist in science popularization and participates in Days of Science in Ukraine and an expert in NGO "We can".

More information at https://www.linkedin.com/in/liubov-zharova-909a1a93/

Oleksandr Artemenko graduated from Ukrainian Academy of Banking of the National Bank of Ukraine (M.S in "Banking"). Currently works as researcher at Young Scientist Research project on the topic "Economic-mathematical modeling of the mechanism for restoring public trust in the financial sector: a guarantee of economic security of Ukraine" (registration number 0117U003924) at the Department of Finance, Banking and Insurance at Sumy State University (Ukraine). Scientific interests include banking, business models, decision-making, trust.

Anna Dyachenko M.S. in "Finance" Ukrainian Academy of Banking of the National Bank of Ukraine. Current position. Post-graduate student, Department of Economics and Business-Administration, Sumy State University. Scientific interests: potential of enterprise, sustainable development, business processes, market.

Olena Obod is a PhD student in "Finance, Banking and Insurance". Research interests include: venture funding for banks, financial stability of the banks, finance analytics, blockchain, cryptocurrency.

Polina Rodionova is a PhD student in "Finance, Banking and Insurance". Research interests include: investments and innovations of the enterprises, venture funding and its drivers, venture tool-box for finance market etc.

About the Editors

Prof. Medani P. Bhandari completed his M.A. in Anthropology (Nepal), M.Sc. Environmental System Monitoring and Analysis (the Netherlands), M.A. Sustainable International Development (USA), M.A. and Ph.D. in Sociology (USA). He is dedicated to conservation of nature and natural resources and social empowerment through research and action project. His purpose of life is to give or contribute to the society fullest through whatever he has, earned, learned or experienced, within the Principles of "BASHUDAIVAKU-TUMBAKKAM and Live and Let Other Live". He has worked with various organizations as consultant – UNEP/Adelaide University, UNDP, IUCN, WWF, WRI, Winrock International, the Japan Environment Education Forum, and the Pajaro Jai Foundation (PJF), along with others. During 2015–17, he served as a Professor of Natural Resources and Environment at the Arabian Gulf University, Bahrain. Prof. Bhandari has spent most of his career focusing on the Sociological Theories; Environmental Sustainability; Social Inclusion, Climate Change Mitigation and Adaptation; Environmental Health Hazard; Environmental Management; Social Innovation; Developing along the way expertise in Global and International Environmental Politics, Environmental Institutions and Natural Resources Governance; Climate Change Policy and Implementation, Environmental Justice, Sustainable Development; Theory of Natural Resources Governance; Impact Evaluation of Rural Livelihood; International Organizations; Public/Social Policy; The Non-Profit Sector; Low Carbon Mechanism; Good Governance; Climate Adaptation; REDD Plus; Carbon Financing; Green Economy and Renewable Energy; Nature, Culture and Power. Prof. Bhandari's major teaching and research specialties include: Sociological Theories and Practices; Environmental Health; Social and Environmental research methods; Social and Environmental Innovation; Social and Environmental policies; Climate Change Mitigation and Adaptation; International Environmental Governance; Green Economy; Sustainability and assessment of the Economic, Social and Environmental impacts on society and nature. In brief, Prof. Bhandari has sound theoretical and practical knowledge in social science and environment science. His field experience spans across Asia, Africa, the North America,

Western Europe, Post-Soviet Countries, Australia, Japan and the Middle East. Professor Bhandari has published 4 books (his recent book is Medani P. Bhandari (2018). Green Web-II: Standards and Perspectives from the IUCN, Published, sold and distributed by: River Publishers, Denmark/the Netherlands ISBN: 978-87-70220-12-5 (Hardback) 978-87-70220-11-8 (eBook), and more than 70 scholarly papers in international scientific journals. Currently, he is serving as a Professor of Inter-Disciplinary Department – Natural Resource & Environment/Sustainability Studies, at the Akamai University, USA and Professor of the Department of Finance and Entrepreneurship, Sumy State University (SSU) Ukraine; and International Program Coordinator, Atlantic State Legal Foundation, NY, USA (remotely).

Shvindina Hanna graduated from Sumy State University (M.S in "Management of Organizations, Ph.D. in "Environmental Economics and Natural Resources Protection"), Post-Doc researcher (University of Montpellier, Purdue University) in "Business Studies and Management", Fulbright Alumni (2018–2019), won ERASMUS MUNDUS scholarships (2014, 2015). Currently works as Associate Professor at the Department of Management at Sumy State University (Ukraine). She is member of CENA community and Researchers' excellence network (RENET). She is member of editorial boards of several scientific journals (Ukraine, Poland, Switzerland), invited reviewer for the international conferences (Poland, Germany, Ukraine). She is author of more than 90 papers. Scientific interests include Coopetition Paradox, Strategies and Innovations, Change Management, Organizational Development, Civic Education and Leadership. She is activist and is a CEO at NGO "Lifelong learning centre", where provide courses on Leadership and Communications, organize the program on empowerment for local community.

More information at https://www.linkedin.com/in/hanna-shvindina-aa39 0924/